# MICROWAVE FOODS:
# NEW PRODUCT DEVELOPMENT

# PUBLICATIONS IN FOOD SCIENCE AND NUTRITION

### Journals

JOURNAL OF RAPID METHODS AND AUTOMATION IN MICROBIOLOGY, D.Y.C. Fung and M.C. Goldschmidt
JOURNAL OF MUSCLE FOODS, N.G. Marriott and G.J. Flick, Jr.
JOURNAL OF SENSORY STUDIES, M.C. Gacula, Jr.
JOURNAL OF FOOD SERVICE SYSTEMS, O.P. Snyder, Jr.
JOURNAL OF FOOD BIOCHEMISTRY, J.R. Whitaker, N.F. Haard and H. Swaisgood
JOURNAL OF FOOD PROCESS ENGINEERING, D.R. Heldman and R.P. Singh
JOURNAL OF FOOD PROCESSING AND PRESERVATION, D.B. Lund
JOURNAL OF FOOD QUALITY, R.L. Shewfelt
JOURNAL OF FOOD SAFETY, T.J. Montville and A.J. Miller
JOURNAL OF TEXTURE STUDIES, M.C. Bourne and P. Sherman

### Books

MICROWAVE FOODS: NEW PRODUCT DEVELOPMENT, R.V. Decareau
DESIGN AND ANALYSIS OF SENSORY OPTIMIZATION, M.C. Gacula, Jr.
NUTRIENT ADDITIONS TO FOOD, J.C. Bauernfeind and P.A. Lachance
NITRITE-CURED MEAT, R.G. Cassens
THE POTENTIAL FOR NUTRITIONAL MODULATION OF THE AGING PROCESSES, D.K. Ingram et al.
CONTROLLED/MODIFIED ATMOSPHERE/VACUUM PACKAGING OF FOODS, A.L. Brody
NUTRITIONAL STATUS ASSESSMENT OF THE INDIVIDUAL, G.E. Livingston
QUALITY ASSURANCE OF FOODS, J.E. Stauffer
THE SCIENCE OF MEAT AND MEAT PRODUCTS, 3RD ED., J.F. Price and B.S. Schweigert
HANDBOOK OF FOOD COLORANT PATENTS, F.J. Francis
ROLE OF CHEMISTRY IN THE QUALITY OF PROCESSED FOODS, O.R. Fennema, W.H. Chang and C.Y. Lii
NEW DIRECTIONS FOR PRODUCT TESTING AND SENSORY ANALYSIS OF FOODS, H.R. Moskowitz
PRODUCT TESTING AND SENSORY EVALUATION OF FOODS, H.R. Moskowitz
ENVIRONMENTAL ASPECTS OF CANCER: ROLE OF MACRO AND MICRO COMPONENTS OF FOODS, E.L. Wynder et al.
FOOD PRODUCT DEVELOPMENT IN IMPLEMENTING DIETARY GUIDELINES, G.E. Livingston, R.J. Moshy, and C.M. Chang
SHELF-LIFE DATING OF FOODS, T.P. Labuza
ANTINUTRIENTS AND NATURAL TOXICANTS IN FOOD, R.L. Ory
UTILIZATION OF PROTEIN RESOURCES, D.W. Stanley et al.
VITAMIN $B_6$: METABOLISM AND ROLE IN GROWTH, G.P. Tryfiates
POSTHARVEST BIOLOGY AND BIOTECHNOLOGY, H.O. Hultin and M. Milner

### Newsletters

MICROWAVES AND FOOD, R.V. Decareau
FOOD INDUSTRY REPORT, G.C. Melson
FOOD, NUTRITION AND HEALTH, P.A. Lachance and M.C. Fisher
FOOD PACKAGING AND LABELING, S. Sacharow

# MICROWAVE FOODS: NEW PRODUCT DEVELOPMENT

*by*

*Robert V. Decareau, Ph.D.*

MICROWAVE CONSULTING SERVICES
AMHERST, NEW HAMPSHIRE 03031

**FOOD & NUTRITION PRESS, INC.
TRUMBULL, CONNECTICUT 06611 USA**

Copyright © 1992 by
# FOOD & NUTRITION PRESS, INC.
*Trumbull, Connecticut 06611 USA*

All rights reserved. No part of this publication may be reproduced, stored in a retrieval system or transmitted in any form or by any means: electronic, electrostatic, magnetic tape, mechanical, photocopying, recording or otherwise, without permission in writing from the publisher.

*Library of Congress Catalog Card Number: 91-77222*
*ISBN: 0-917678-30-3*

*Printed in the United States of America*

# PREFACE

The original concept for this book goes back almost ten years when the microwave oven was enjoying a sales boom. This was before too many microwavable products had appeared in the market place. So much new material was being generated that the book became a sort of living document with no completion date in sight. Eventually, it became evident that unless a manuscript was submitted soon, it would be of little use to the food product development community.

The book was designed to provide the reader with some of the history of the microwave oven in food service that this might, by indicating what has been tried before, suggest other opportunities for new product development. It was also intended as a not too technical treatment, since the product developer will have had the necessary education in food science and skills in product development to recognize the microwave oven as another heating method that requires relatively minor adjustments in technique to fulfill his or her responsibilities. It would be presumptuous to think that it would be more than a reference book to keep handy when planning microwavable product development efforts. The writer sincerely hopes that the reader will recognize the tremendous opportunities that exist for microwavable food product development and foodservice.

The first chapter on the history of the microwave oven was meant in part to record for future generations its development to date. It is an interesting story and one in which only a few of the original participants remain. Chapter two briefly discusses microwave heating technology and factors that affect microwave heating results. This may be the most important chapter for food product developers to become familiar with. Chapter three describes microwave oven features, standardization and safety matters; factors that the product developer needs to be familiar with in order to write proper directions for heating microwavable products. Chapter four on packaging may be the second most important chapter since the package has a great deal of influence on microwave heating results. Chapter five on browning is important to the developer in learning to address this problem. Chapter six by a discussion of what others have done in product development perhaps will save the developer from making some of the mistakes of others and also save reinventing the wheel. This chapter does not provide all the answers to new product development, but hopefully will be a useful guide. Chapter seven is a review of the effect of microwave on nutrients in food, which should indicate to the developer that there is a potentially good selling point in being able to promote the fact that microwave heating and cooking are not harmful to nutrients. And finally, on microbiological matters. Chapter eight relates what others have reported on the effect of microwave heating on microorganisms in food. It should be clear that, when properly applied, microwave heating can deliver a safe product.

The author is grateful to his many friends and colleagues in food science and technology who have not discouraged this effort. He is also grateful to his publisher, John O'Neil, for his patience and gentle but persistent prodding over at least a half dozen years that eventually resulted in a completed manuscript.

ROBERT V. DECAREAU

# CONTENTS

| CHAPTER | PAGE |
|---|---|
| 1. HISTORY OF THE MICROWAVE OVEN | 1 |
| 2. FUNDAMENTALS OF MICROWAVE HEATING | 47 |
| 3. THE MICROWAVE OVEN | 67 |
| 4. PACKAGING FOOD PRODUCTS FOR THE MICROWAVE OVEN | 87 |
| 5. BROWNING AND THE MICROWAVE OVEN | 117 |
| 6. NEW PRODUCT DEVELOPMENT FOR THE CONSUMER MICROWAVE OVEN MARKET | 129 |
| 7. NUTRITION | 165 |
| 8. MICROBIOLOGICAL CONSIDERATIONS | 189 |
| EPILOGUE | 203 |
| APPENDIX | 207 |
| INDEX | 209 |

# CHAPTER ONE

# HISTORY OF THE MICROWAVE OVEN

## INTRODUCTION

Knowledge of a market is essential to the development of products for that market. In the case of the microwave oven, this is an appliance unlike any other appliance because it heats food throughout rather than just at the surface and as a result it opens the door to a wealth of new food products developed specifically for use in it. Indeed, microwavable foods may one day represent the largest prepared food market of all. Why this may be so will become abundantly clear from the discussion in the following pages of the development of the microwave oven, its broad possibilities in the home, in foodservice in general, and its proliferation. Eventually, some time in the near future, the microwave oven will become commonplace throughout the civilized world. Already in 1991 various sources indicate that 70 to 80% of United States households are using microwave ovens. The estimates for some other countries are: Australia—43%; Canada—41%; Japan—44%; New Zealand—29%; and The United Kingdom—28%. These percentages are increasing rapidly. Sales of microwave ovens in western, central and northern Europe are increasing so rapidly that market saturation is expected to attain by 1992 a level equal to that of the United States in 1989. It has been said that in some parts of Africa the conventional stove will be by-passed as the microwave oven will transition directly from the open fire.

Not since the discovery of fire in deep pre-history has a phenomenon had such a striking influence on life styles as microwave cooking and heating. The effect on humanity is likely to continue for decades, perhaps centuries.

The microwave oven was introduced to the public shortly after World War II, and within a few short years had earned a degree of respectability in a number of commercial foodservice applications. It was recognized early that the microwave oven and frozen foods were natural partners. Quite some time later in 1970, Mr. E. W. Williams, then publisher of Quick Frozen Foods magazine, commented that the microwave oven would be one of the greatest motivating forces the frozen food industry ever had. He added that the potential was for 10 million microwave ovens in homes in the next decade (Williams, 1970). His forecast was only slightly low. The progress of consumer microwave oven sales is shown in Table 1.1.

TABLE 1.1
MICROWAVE OVEN SALES: 1970 TO 1990

| Year | Sales |
|---|---|
| 1970 | 40,000 |
| 1971 | 100,000 |
| 1972 | 300,000 |
| 1973 | 440,000 |
| 1974 | 725,000 |
| 1975 | 1,000,000 |
| 1976 | 1,600,000 |
| 1977 | 2,150,000 |
| 1978 | 2,501,000 |
| 1979 | 2,807,000 |
| 1980 | 3,608,000 |
| 1981 | 4,422,000 |
| 1982 | 4,071,000 |
| 1983 | 5,933,000 |
| 1984 | 9,132,000 |
| 1985 | 10,883,000 |
| 1986 | 12,444,000 |
| 1987 | 12,610,000 |
| 1988 | 10,988,000 |
| 1989 | 10,598,000 |
| 1990 | 8,126,000 |
| 1991 (est) | 8,479,000 |

Many of the early microwave foodservice concepts were based on heating precooked frozen foods in microwave ovens. The commercial microwave oven market was pursued first because microwave oven technology had not advanced sufficiently to permit the marketing of an affordable and reliable household version. The financial risks were too great at the time to attempt the introduction of a consumer microwave oven.

The pioneering effort in microwave oven development was carried out at Raytheon Company in Waltham, Massachusetts. This firm introduced the first microwave oven and developed a strong patent position on oven design. But it must have required a tremendous act of faith on the part of Raytheon Company management, or an uncanny recognition of the future potential of this device, to support a development effort that for the better part of two decades could not have been profitable.

## THE COMMERCIAL MICROWAVE OVEN

The commercial microwave oven market encompasses vending of food to be microwave heated, cooking and heating of foods in convenience stores, lounges, taverns, hospitals and other institutions, snack bars, and restaurants of all sizes. This

Fig. 1.1 Early model Radarange™ microwave sandwich heater introduced in 1946 (Courtesy Raytheon Company, Lexington, MA).

market is, and always will be, a small fraction of the size of the consumer oven market in numbers of ovens as well as in dollar value.

The idea of an electronics firm entering the restaurant equipment business in the early 1950s must have appeared to most observers as pure folly. Today, with the 20/20 vision of hindsight, this decision might be described as pure genius as the firm that introduced the microwave oven under the catchy name "Radarange"™ eventually realized a substantial profit in the appliance business. Today, most consumer microwave ovens and a good percentage of the commercial microwave ovens are manufactured off-shore.

The Radarange™ microwave oven (Fig 1.1) was introduced in October 1946 at a special showing in New York City to food editors, home economists, restauranteurs, and airline operators (Anon. 1946). Aware of the limitations of this sandwich heater Raytheon already had designed a more practical unit (Fig. 1.2) which was intro-

Fig. 1.2 Model 1130 Series Radarange™ microwave oven became available in 1947 (Courtesy of Raytheon Company, Lexington, MA).

duced in a very tentative and cautious manner, to a few restaurants in the New England area (Anon, 1947a). This almost immediately brought about the recognition that there were important differences between microwave and conventional cooking. Much technology in food preparation had to be developed to support the marketing of this new, seemingly incredible, way of cooking.

Early research was supported by a grant from Raytheon Company to the Food Science Department at the Massachussetts Institute of Technology in Cambridge, Massachusetts at that time under the chairmanship of Professor Bernard E. Proctor. The results of that research may be found in the scientific literature (Proctor and Goldblith, 1948, 1951). Other work was carried out at the U.S. Army Food and Container Institute in Chicago (Bollman et al, 1948) and by the U.S. Navy (Sussman, 1947) where its use aboard submarines was contemplated.

Based on the conviction that the microwave oven offered great sales potential, but required additional technological support in the food preparation area, Raytheon Company established a food laboratory in 1953 managed by Dr. David A. Copson, who at the time was a recent graduate of the Massachusetts Institute of Technology and one of Professor Proctor's graduate students. After little more than a year of the laboratory's existence, the company sponsored the first microwave oven seminar in November, 1954 to report on the results of its research in microwave cooking (Anon, 1954).

A considerable body of food research was carried out in those early days using microwave ovens that were little more than engineering prototypes. Some good basic microwave cooking know-how was generated that still stands in good stead today, more then 40 years later. It was also apparent to those foodservice operators who purchased the first microwave ovens that the equipment and the food preparation techniques had some serious shortcomings, not the least of which was the nonuniform cooking pattern of the oven, how to cope with such rapid cooking and what to do about the lack of surface browning.

The first seminar presented the Raytheon Company Food Laboratory's experimental results and the experimental results of other groups; academic, commercial and government. Those in attendance at the seminar included representatives from Howard Johnson's and Stouffer's restaurants, Corning Glass Company, Cornell University, Massachusetts Institute of Technology, Columbia University, Michigan State University, Simmons College, The U.S. Army Quartermaster Food and Container Institute, The U.S. Navy Commissary Research Facility, Detroit Edison Company, The National Live Stock and Meat Board, Tappan Stove Company, and Frigidaire Division of General Motors Corporation.

In 1955, Raytheon Company established a sales force of equipment salesmen backed up by a professional chef. The Food Laboratory worked closely with this team to refine microwave cooking techniques in support of the sales effort. Slowly and steadily sales in the restaurant field increased. New and improved microwave ovens were introduced in 1957 and in succeeding years. Some of the early commercial Radarange™ microwave ovens are shown in Fig. 1.3 to 1.8.

Other equipment manufacturers began to show interest. The Bruder Corporation (later to become a part of Litton Industries) with their Heat 'N Eat microwave oven equipped with a push button timer became a serious contender for the hot food vending market (Fig. 1.9). About this same time General Electric Company entered the commercial microwave oven market with ovens designed to be stacked to conserve floor space. The first conveyorized microwave oven for foodservice was shown by the Philips Company of Holland at the Rotterdam "Floriade" exhibition in the Summer of 1960 and again in 1961 at the ANUGA exhibition in Cologne, Germany. Preplated meals were heated on a continuous basis from $-25$ C to 80 C at a rate of 150 to 200 meals per hour (Püschner, 1966). Although an apparent technical success, efforts to market this equipment in the United States as well as in Europe were not successful.

Fig. 1.3 Model 1161 Radarange™ microwave oven, circa 1954. This oven was equipped with two, 800 watt magnetron microwave generators directly inserted through the roof of the oven cavity. The oven could be operated at half power by switching off one of the magnetrons (Courtesy of Raytheon Company, Lexington, MA).

Commercial microwave oven sales continued at a steady but still disappointing rate until a breakthrough occurred in 1966 when Litton Industries Atherton Division in Palo Alto, California introduced a 1 kW counter-top microwave oven (Fig. 1.10) at a price under $1000 (Moore, 1966). A few years later, Raytheon Company transfered its commercial microwave oven business to Amana Refrigeration, which became a subsidiary in the 1970s. Sometime in the 1970s Japanese oven manufacturers, Sharp and Panasonic, introduced commercial ovens into the United States. Sharp Corporation had begun production of microwave ovens as early as 1962. The Philips Company of Eindhoven, Holland, and Husqvarna of Huskvarna, Sweden also began to market commercial microwave ovens in Europe. And there were some efforts to market their ovens here in the United States.

There also was some activity in the development and marketing of combination microwave forced convection ovens. Forced hot air convection ovens were first marketed in the late 1960s and grew rapidly in popularity because of the considerably shorter cooking times made possible by moving hot air rapidly over food surfaces.

Fig. 1.4 Model 1170 Radarange™ microwave oven circa 1954. This counter top microwave oven was equipped with a single 800 watt magnetron directly inserted through the roof of the oven cavity. Several power settings were possible (Courtesy of Raytheon Company, Lexington, MA).

The combination of this technique with microwave heating seemed a natural progression in cooking technology to give typical browning results in microwave time. The initial effort to produce and market such an oven occured in England where the Hirst Corporation introduced the Micro-Aire oven (Fig. 1.11) (Constable, 1973; 1975). These ovens were normally operated at 400–500 F. Microwave power was variable from a few hundred watts to about 2.5 kW at 2450 MHz. Marketing efforts were not too successful in the early years of this oven's existence, however in recent years its adoption by British Rail for use in the galleys of its passenger trains gave it new life. Later, Litton Microwave Cooking introduced its Jet-Wave model forced convection microwave oven (Fig. 1.12) to the foodservice market.

Fig. 1.5 Mark III Radarange™ microwave oven, an improved version of the Model 1161, circa 1957 (Courtesy of Raytheon Company, Lexington, MA).

## FOODSERVICE APPLICATIONS

Although the microwave oven originally was promoted as a cooking tool, it soon was evident that its use as an element of foodservice systems would be the more promising application. Thus, instead of becoming a prime cooking device, which for obvious reasons it was ill-suited, it found a more useful role as an expediter; that is, to finish off various dishes that had been partially prepared by other cooking methods. Examples of this role included: quick thawing of frozen meat, fish and poultry items; finishing off preseared steaks and chops; heating portions of refrigerated precooked foods; occasionally cooking a baked potato or an ear of corn

Fig. 1.6 Mark IV Radarange™ microwave oven replaced the Model 1170. A unique vertical sliding door eliminated the lost space of a pull down door in crowded restaurant kitchens. Courtesy of Raytheon Company, Lexington, MA.

and so forth. Some examples of microwave foodservice systems are discussed in the following pages.

**Restaurant foodservice**

The concept of a microwave restaurant backed up by a refrigerator and a freezer was another foodservice application conceived well before its time. Indeed, it may just be wishful thinking that such a concept could be realized, but there have been several attempts to do so. The first known attempt was by the White Tower restaurant chain in Boston, Massachusetts. This restaurant chain installed one of the very first microwave ovens (Anon, 1947a). The first menu items were raw hamburger and onion on a roll cooked in just 30 seconds, and frankfurters in a bun cooked in 10 seconds. These short cooking times were possible because the oven power of these first ovens was close to 2 kW. The record does not show how well these products

Fig. 1.7 Mark V Radarange™ microwave oven replaced the Mark III. It featured a countertop design and a vertical sliding door (Courtesy of Raytheon Company, Lexington, MA).

were received by customers, but the thrill of seeing food heated in this way must have been worth the cost of a hamburger or a hot dog.

Thompson's Spa, at one time a popular Boston chain of eleven restaurants and a division of the Sheraton Corporation of America, was another of the very first to consider a foodservice system based on microwave heating of frozen food items. As envisioned, the system would have completely eliminated waste because a meal would not have been microwave heated until it was ordered and the use of plastic plates would have eliminated dishwashing (Anon, 1947b).

One of the first highly successful applications of the microwave oven was at the Green Ridge Turkey Farm restaurant in Nashua, New Hampshire beginning about 1954. This restaurant at the time served over 100 tons of turkey per year. With the advent of the microwave oven, the restaurant modified its turkey dinner operation.

Fig. 1.8 Mark VI Radarange™ microwave oven was a small lightweight oven designed to meet the needs of the vending industry (Courtesy of Raytheon Company, Lexington, MA).

Turkeys were now roasted in advance, cooled, sliced and exact portions weighed and stored in the refrigerator on sheet pans. To process an order, a scoop of turkey dressing was placed on a dinner plate, covered with a portion of turkey meat, brushed with turkey stock to moisten the meat, the plate heated for about 15 seconds in a 1.6 kW microwave oven and the remaining components of the meal added from the steam table. Thus in less than a minute an order was filled and ready to serve. The main advantage of this system, in addition to speed, was in the better utilization of the turkey by slicing and portioning the meat cold rather than hot. This resulted in several more portions per turkey. In addition the labor requirement was reduced by at least two cooks. This alone resulted in a very rapid return on the investment in microwave ovens used in the system (Anon, 1968a).

Another restaurant owner claimed that improved portion control and the elimination of the warming oven and steam table shrinkage permitted the same daily volume to be handled with 60 per cent fewer turkeys. The 700 seat Devon Gables Tea Room in Bloomfield, Michigan used its first microwave oven to build its lobster business from virtually nothing to over 150 orders on a typical Sunday. Although the ovens were scheduled primarily for lobster tails and fish, they provided the kitchen with

Fig. 1.9 Bruder Corp. Heat 'N Eat™ microwave oven circa 1963 was sold mainly to the food vending trade (Courtesy Litton Systems, Inc., Beverly Hills, CA).

greater flexibility when faced with the unexpected. For example, a baked potato was ready in three minutes and steaks and chops could be defrosted in less than one minute (Anon, 1968a).

The Covered Wagon Inn in Innkster, Michigan, a 450-seat restaurant found that the microwave oven allowed an additional 125 to 200 people to be accomodated on a busy Saturday night. Almost all of the entrees such as steaks, chops, fish and lobster were kept frozen then thawed in the microwave oven as needed. This improved food control and reduced food costs 3 to 5 per cent.

Many other examples of successful microwave oven applications in the early years of its availability could be cited, but these few serve to illustrate how its use in a systematic manner could bring order to foodservice operations.

In summary the main advantages of microwave ovens in restaurants are:

- Better portion control
- Reduced labor costs
- Increased menu variety as a result of the thawing ability of the ovens
- Faster seat turnover
- Hotter, tastier, juicier foods
- Food costs and inventory more easily controlled.

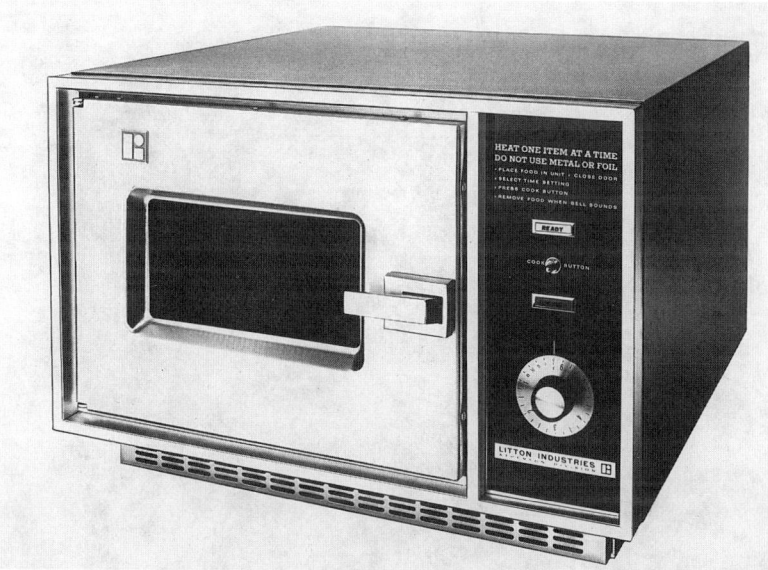

Fig. 1.10 Litton Model 500 the first truly compact counter top microwave oven with one kilowatt of microwave power (Courtesy of Litton Systems, Inc., Beverly Hills, CA).

One of the most noteworthy restaurants and the one that received the most publicity in the trade press was an experimental restaurant on 42nd Street near 5th Avenue in New York City (Anon, 1963a). In this self-service type restaurant the customer could choose from some 30 to 40 preplated meals. The meals were stored in refrigerated display cases where attractive visuals assisted customers in making their selections. Salads, rolls, desserts and beverages were selected from a cafeteria line. After payment was made to the cashier, the customer proceeded to the dining area where each table was equipped with two microwave ovens. The customer placed his plated meal with its shrink film cover into one of the ovens and closed the oven door. The oven automatically cycled on for a period of time based on the weight of the plated meal and an initial temperature of 38 to 40 F. Thirty to 90 seconds later a light came on to announce that heating was completed.

All the meals in this restaurant were prepared, plated, lidded and blast frozen in a basement kitchen below the restaurant and stored at −10 F. Meals were transferred to the refrigerated display cases to thaw prior to business hours. One of the problems observed in this operation was a variation in the initial temperature of the meals; that is some meals were not completely thawed. Thus customer's meals heated on the basis of weight from an assumed refrigerated condition often were insufficiently heated, though repeat customers learned to return the plate to the oven for additional heating. Many customers found this confusing and did not return. Although

Fig. 1.11 Hirst Micro-Aire combination forced convection and microwave oven (Courtesy Mealstream Ltd, England).

operated at this prime New York City location for about a year, management judged that the concept required further development and decided to discontinue the experiment.

Another experiment in foodservice that deserves mention was the Gingham Kitchen snack bar. This was an ambitious effort to promote commercial microwave oven sales by the Raytheon Company in the early 1960s by marketing a complete, "packaged" Fast Food, foodservice system. A number of these snack bars were installed in New York City; e.g., at Columbia University and Macy's department store, in 1962. Unlike many of today's fast food operations, the Gingham Kitchen did not have control over its food offerings although a belated effort to do so fell short of success and management decided to stop what at that time was a losing proposition.

The Gingham Kitchen consisted of a prefabricated snack bar, including the counter, back bar and stools, along with all of the foodservice equipment required for its operation; i.e., microwave ovens, coffee makers, beverage dispensers, freezers, refrigerators, et cetera. Food items offered included a pre-assembled frankfurter on a roll, a precooked hamburger on a roll, Danish pastries, muffins, doughnuts,

Fig. 1.12 Litton Jet-Wave microwave oven, the first combination convection and microwave oven built in the U.S. (Courtesy Litton Systems, Inc., Beverly Hills, CA).

frozen thick shakes, coffee and carbonated beverages. No foods were prepared from scratch, therefore no highly skilled foodservice personnel were required.

The thick shake was a unique item, packaged in a tall coated paperboard cup, frozen and microwave thawed on demand. It had variable success mainly because of the lack of temperature control in distribution and storage. When heated from the solidly frozen condition it was an excellent product, one that was cold yet soft enough to be eaten with a spoon, but generally too thick to be consumed by sipping through a straw. Storage temperature changes due to frequent openings of the freezer resulted in texture changes, quality loss, and unpredictable thawing results as a result of ice crystal growth.

The frankfurter and hamburger items were generally unsatisfactory, but could have been successful with today's knowledge of microwave heating technology. Although it would have been ideal to heat all foods from the frozen state, fast food operation traffic is generally too intense to tolerate the extra time required. Instead, programmed tempering of quantities of each item predicted from a knowledge of traffic levels would allow heating from just above the frozen state in about one-half the time; that is, heating cycles of 15 to 30 seconds. It is interesting to contemplate how this early fast food concept might have fared if these products were successfully heated in view of the success of so many fast food operations in recent years where tight control has been exercised over product quality.

## Hospital foodservice

Foodservice is one of the largest single expenses in the hospital operating budget. According to Hartman (1965) salaries and wages account for 60 per cent of the operating costs of hospital food service. Although this ratio may have changed over the years due to rising food and labor costs, foodservice still remains a major hospital operating expense.

Cease (1967) in a presentation before the Society for the Advancement of Food Service Research said, we must find a way of storing labor. One way in which this is being accomplished is through the increased use of precooked foods. The concept of heating precooked meals for hospital patients was an early successful application of the microwave oven. A system was placed in operation at the Kaiser Foundation Hospital in Walnut Creek, California in 1955 (Park, 1957), and later extended to other hospitals in the system. Foods were prepared in quantity several days in advance in the hospital kitchen with the work load spread over the full work day. The food was then frozen or refrigerated depending on when it would be used. Meals were assembled from cold food onto china plates, covered with a transparent film and stored on shelves adjacent to the microwave ovens. Soups also were dished up at this time, then one plate of food and one cup of soup were heated together in each of two microwave ovens in 40 to 50 seconds and placed in the hot section of a tray cart. This short heating time was possible because of the high power level (1.6 kW) of these ovens. The cart was transported to the ward where the hot plate of food and soup were placed on the appropriate trays and delivered to the patients.

Other hospitals adopted microwave ovens following the Kaiser Foundation example. Labor savings of more than $56,000 were claimed by one hospital using precooked frozen foods purchased from outside suppliers and heated in microwave ovens (van Gemert, 1965). Another claimed substantial labor reductions as well as an 18 per cent savings in basic food costs (Anon, 1966a). Another Kaiser Foundation hospital was said to have saved $1200 per month while serving gourmet meals to patients from essentially a kitchenless system (Anon, 1966b).

In some hospitals plated meals were heated at the ward level just before serving to insure a truly hot meal for the patient. At St. Vincent's hospital in Toledo, Ohio a microwave oven accompanied the food cart to the wards when their system was first introduced. The Walter Reed Army hospital in Bethesda, Maryland had microwave ovens mounted on the tray carts for heating meals just before delivery to patients. Others located the microwave ovens in the ward pantries and deliver chilled plated meals to the pantries for heating just prior to serving.

It was some years before the food industry offered preplated frozen meals to the hospital foodservice market. Industry received encouragement from the fact that some hospitals were converting their kitchens into production centers where foods would be prepared in quantity for future use. Meal reheating in microwave ovens in the ward pantry replaced central meal heating in many hospitals although the advent of insulated carriers and other heat maintenance systems in some cases returned meal heating to kitchens. The microwave oven reheating capability was recognized by hospital foodservice management as a means to use labor more efficiently and accordingly to reduce labor costs. Of equal importance, hot, nutritious and tasty meals would be appreciated by patients.

Continuous meal heating ovens for hospital foodservice were introduced in the late 1960s by Automatic Food Supply AB, Växjö, Sweden (Rejler, 1970). The oven was only one element of a system that included storage magazines in refrigerated and/or freezer storage spaces to dispense individually packaged meal components, a pneumatic delivery system to speed selections to appropriate hospital wards, and a control console located close to the point of service (e.g., in each ward pantry).

In concept, a foodservice employee made selections indicated on the patient's menu card by pressing the appropriate buttons on the control console. The selections were automatically dispensed from the storage unit onto conveyor belts that carried the selections to be heated through a microwave oven at a rate of one every six to eight seconds then via the pneumatic delivery system to the ward pantry. Salads and other cold foods and beverages were carried on a separate belt that by-passed the microwave heating system. The items then were assembled on a tray at a rate of five to six trays per minute for delivery to patients. The advantages of pneumatic delivery may never be realized as only a demonstration unit was built.

In an installation of the system in the United States at West Jersey Hospital, Camden, New Jersey, as described by Brown and Doyon (1973), the kitchen was operated as a production facility, five days a week. The precooked foods were portioned into heavy duty polyethylene containers, 4-in by 5-in by 1 3/8-in: one container with a single compartment for soups, entrees, salads and desserts; the other a divided container for vegetables. The containers were heat sealed with a clear polyester film and quick frozen in a liquid nitrogen freezer to $-10$ F ($-23.3$ C).

At West Jersey Hospital, the system worked as follows. About two days before specific frozen meal items were to be used, they were transferred to a 35–40 F (1.67–4.44 C) tempering room and loaded into heavy duty plastic holders (cassettes) each with a capacity of 20 to 28 portions. Prior to meal time the cassettes were fitted into dispensing magazines of which there were 24 in the refrigerated storage space. Each magazine could accommodate up to nine cassettes or a minimum of 180 portions. Twelve magazines were assigned to foods to be heated and twelve to foods to be served cold. The magazines were operated from the remote control console located a short distance away in the kitchen assembly area. Selections were made by the operator who pressed the console buttons that corresponded to items on the menu card then pressed the Delivery button. As mentioned before those items to be heated passed through a microwave conveyor oven while the cold foods were delivered by a conveyor that by-passed the microwave oven on route to the collating point. The completed tray was covered with an insulated lid and placed in a tray cart for delivery to the wards. These conveyor microwave ovens had sufficient power to deliver heated selections at a rate of one every six to eight seconds. Trays were assembled at a rate of five to six per minute.

The advantages claimed by Brown and Doyon (1973) for the system at West Jersey Hospital were:

- A 40-hour work week instead of 12-hour shifts each day.
- Flattened peak work periods by separating production from service.

- Extended product life by packaging and freezing.
- Waste reduction.
- Rapid heating retained palatability.
- Labor reduced 46 per cent.
- Tray assembly labor reduced 80 per cent.
- Dishwashing time reduced 75 per cent.

In 1973, the Finessa Corporation, a Swiss firm added a microwave tunnel oven to its line of hospital food service equipment. In the Finessa system, foods were cooked, portioned onto china service, and covered with a color coded plastic cover. The meals were chilled to 34 to 36 F (1.11–2.22 C) and heated in the 15 kW tunnel oven at a rate of 400 14-ounce meals per hour. When meals were plated on plastic service the rate increased to 615 meals per hour, thus showing the competition that china service offered for the available energy. The Finessa tunnel oven has been used for school lunch programs, in-plant feeding and cafeteria service in Switzerland, West Germany, The Netherlands and Spain. For in-plant foodservice, hot foods were plated into disposable compartmented trays that were automatically lidded with a film cover. The trays were topped off with respect to temperature by passing them through a microwave tunnel oven, after which they were packed in insulated carriers for transporting to the plant where they were distributed to workers.

The most recent addition of continuous microwave heating equipment to the market was the unit designed by and built by Enersyst, Inc., Dallas, Texas specifically for hospital foodservice. Called the Food Finisher (Fig. 1.13), this system's special feature that made it different from all others was the simultaneous application of humidified air during the early part of the microwave heating cycle. The humidified air was directed down onto the uncovered plated food through small openings in air ducts above the conveyor that gave the air direction and much more effective heat transfer. The use of moist warm air in this manner prevented dehydration while improving the uniformity of heating. Hot air provided from below in the first section of the oven heated the plate thus reducing heat loss from the food to the plate. Several different heating programs may be applied depending on the size and weight of the preplated meal. The preplated meals are heated at a rate of 300 meals per hour. As they leave the oven they are placed in insulated trays, covered and distributed to the wards. A second generation version of the Food Finisher used an infrared sensing system for monitoring surface temperature. This information was used in a feedback circuit to control the heating program.

## Cafeteria feeding

The first example of self-service microwave heating of prepared meals was demonstrated by Raytheon Company in one of its own plant cafeterias (Anon, 1961). Eleven of the company's cafeterias were using this system in 1961. A variety of meals were prepared and plated by a local caterer on attractive molded pulp plates

Fig 1.13 Enersyst Development Center Meal Finisher continuous system for jet impingement assisted microwave heating of hospital diets (Courtesy Enersyst Development Center, Dallas, Texas).

(Chinet by Keyes Fibre Co.), overwrapped with a shrink film and frozen. Dinners were transferred to refrigerators 24 hours in advance to allow them to thaw. Several microwave ovens with color coded push button timers were located against one wall of the cafeteria. Customer using the facility selected their meals from a refrigerated display case and heated them according to directions on the cover by placing their meal selection in one of the ovens and pressing the appropriate timer button. Heating time in 1.6 kW microwave ovens averaged about one minute per meal.

A cafeteria system in Sweden associated with highway refueling stations used a conveyorized microwave oven built by the Husqvarna firm of Huskvarna, Sweden. Patrons made their selections from a menu board on entering. The selections were then plated in the kitchen from chilled prepared foods, covered and passed through a conveyor microwave oven to heat. They were then garnished and conveyed to the serving line.

Husqvarna conveyor ovens have also been used in school lunch programs in which precooked entrees, packaged in polyethylene coated paperboard cartons, were heated in a narrow tunnel oven located on the serving line. Students picked up a plate of potatoes or other starch item, fruit and beverage from the cafeteria line, then finally their hot entree as it left the oven (Fig. 1.14).

Fig. 1.14 Continuous meal heating system used for school lunch feeding. Heated meals exit from the rectangular opening along the edge of the serving line (Courtesy Huskvarna, Sweden).

**Food vending**

Hot sandwich vending was one of the very first microwave foodservice concepts to be evaluated. The broad variety of hot sandwiches and their general popularity suggests that this is a product for which a concerted effort should be made to develop a successful heating and vending method. Sandwich heating is not such a lengthy process that a microwave heated and vended sandwich would not be a successful application. There are many locations where the traffic volume is not so heavy that a 15 to 30-second vending cycle would create a bottleneck and discourage customers.

A prototype microwave sandwich vending machine was field tested in the mid-1950s in Boston, Massachusetts. The machine described in the patent (Torsiello et al, 1968) vended a selection of four sandwiches heated from the refrigerated state by means of an integral microwave oven. The sandwiches were wrapped in cellophane and consisted of bread or a roll with a meat filling. The quality of the product suffered mainly because the roll or bread component became tough and leathery during heating.

The use of microwave ovens to heat foods obtained from banks of food vending machines in locations such as office buildings, factories, schools, and the like was one of the more successful early applications of these ovens. This activity continues strong today. To remove any question about heating time, a push button timer system was developed so that simple directions could be provided. Thus a hamburger had its own button labelled Hamburger; Casseroles, pastries, etc., had their own preset push buttons labelled accordingly. Bruder Corporation, Cleveland, Ohio, which was acquired by Litton Industries, quickly exploited the food vending market with its Heat'N Eat microwave oven. The oven was manufactured for Bruder Corporation by a private label manufacturer. After acquisition by Litton Industries, Bruder Corporation used the push button version of the Model 500 oven (Figure 1.15). According to Ellis (1983) more than 200,000 microwave ovens were being used by the vending industry. The number of ovens used in this application has increased considerably since then.

Automatic Food Supply demonstrated its microwave heating system, mentioned earlier in the section on hospital foodservice, in a vending application at a small outlet in Växjö, Sweden. During the day the outlet was operated as a restaurant with attendants actuating the system to deliver customer requests. After closing, the system was operated by the patrons through a vending console located outside of the restaurant.

A study (Smith and Harris, 1974) that resulted in a prototype microwave vending machine is well worth relating for the knowledge gained and the potential opportunities suggested.

A requirement existed for around-the-clock, hot meal service for US Air Force flight lines and other remote military locations that could not justify a conventional foodservice operation. A review of commercial vending machines indicated that a

Fig. 1.15 Litton Model 500 PB, a push button version of the Model 500 microwave oven directed specifically to the food vending trade (Courtesy of Litton Systems, Inc., Beverly Hills, CA).

variety of units were available that could dispense cold food items. Aside from canned food vending machines that maintained the cans hot until dispensed, and microwave ovens used in conjunction with refrigerated vending machines that dispensed some preplated meals and sandwiches, no equipment was available that could handle this hot food service requirement in a satisfactory manner.

The study was carried out to determine if a practical device could be built to heat individual and groups of packages of prepared frozen food items to their appropriate service temperatures in two minutes or less based on a maximum of 12 ounces of food. These specifications were established to identify the dimensions of a storage unit from which one could obtain a hot, 3-component dinner or an individual food item. The time factor was felt to be reasonable for locations with limited traffic. The weight factor was considered reasonable for a frozen 3-component dinner based on offerings being made in the retail frozen dinner market. The study included an evaluation of packaging materials, packaging concepts, heating methods, and a conceptual design of a mechanism to meet the prescribed specifications.

The results of the study were positive. The concept of a device to heat packaged frozen food portions to serving temperature and dispense them in about two minutes was determined to be feasible and that a functional device could be built. Food container materials and package designs also were identified. A workable solution to the problem of heating food products that required a crisp surface to be acceptable also was provided.

Subsequently, a contract was awarded by the U.S. Army Natick Research and Development Laboratories, Natick, Massachusetts for the Design and Fabrication of a Device for Storing, Heating and Dispensing Meal Components. Microwave heating was identified as the only method that could meet the heating time limitations, and an impingement hot air system was selected for use in conjunction with microwave energy for heating foods requiring a crisp exterior. The heating unit consisted of three contiguous microwave ovens each with a microwave power output of 1.0 kW and controlled temperature, up to 550 F., jet impingement hot air heating. Storage stacks were provided in a prefabricated freezer to hold 50 portions each of 12 different food items. All operations from selection to heating and dispensing up to three selections at one time were controlled by a solid state program controller. A schematic drawing of the device is shown in Fig. 1.16.

It is not necessary to go into the details of construction of this unit. A number of other solutions to the mechanism of storage and dispensing will suggest themselves to those interested. Essentially, the components of a dinner are chosen by turning selector switches on the front of the unit to the appropriate positions (Fig. 1.17) and pressing a start button. Subsequently, each selection moves into the contiguous ovens where they are heated acccording to programs developed specifically for each item. The programs were controlled by a reprogrammable control system (5TI Program Controller by Texas Instruments). A typical heating program for an entree item might consist of one minute at full microwave power, followed by one minute at 15 to 85% of full power and completed by up to one minute of continuous heating

# HISTORY OF THE MICROWAVE OVEN

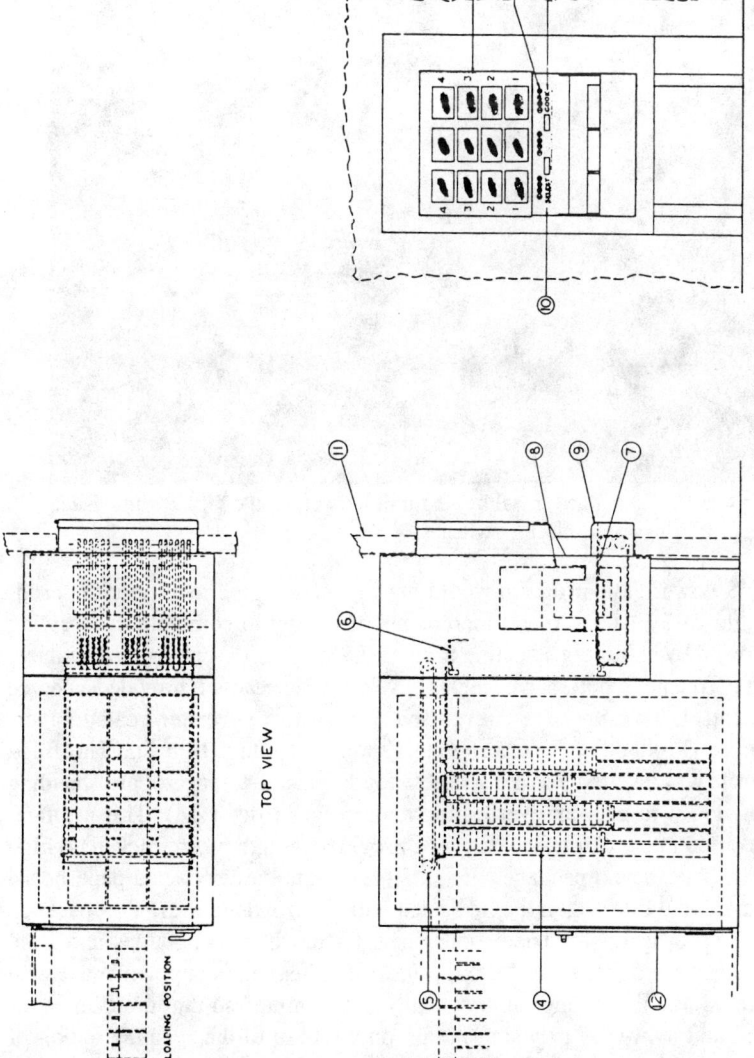

Fig. 1.16 A schematic illustration of the meal heating unit built for the U.S. Air Force by Enersyst Development Center about 1974.

Fig. 1.17 The U.S. Air Force meal heating unit illustrating the facade with visuals of the food offerings and selector switches.

at 15 to 100% power. The program would have been designed to temper the product to just below 32 F, then to add increments of energy to complete the thawing process followed by continuous heating to raise the product to serving temperature.

Packaging played an important role in this development because it provided a degree of heating control. Two packages were used. One was a polyester coated paperboard casserole (Product of American Can Company) with a film lid. The filled lidded casserole was inserted into a disposable foil laminated-to-paperboard shielding device (Falk, 1978) to prevent overheating at the edges (Fig. 1.18). This arrangement was inserted into a chipboard carton to provide strength when stacked 50 high in the freezer. The second package (Fig. 1.19) was a polyester coated paperboard carton with a die-cut opening in the lid sealed with a shrinkable film. This package was designed specifically for fried foods such as french fried potatoes and fried chicken. When these items are selected, microwave heating is supplemented with jet impingement hot air heating, automatically programmed so that the film cover would shrink and split thus exposing the product within to the crisping action of 400 F or higher temperature hot air. The jet impingement technique provides substantially faster heat transfer than undirected hot air and crisps the food simultaneously with internal microwave heating. Upon completion of the heating step the selected items are delivered down a shute to the customer.

An integrated vending machine such as this could be programmed to temper frozen meats, fish, poultry and other foods for customers to take home for immediate use;

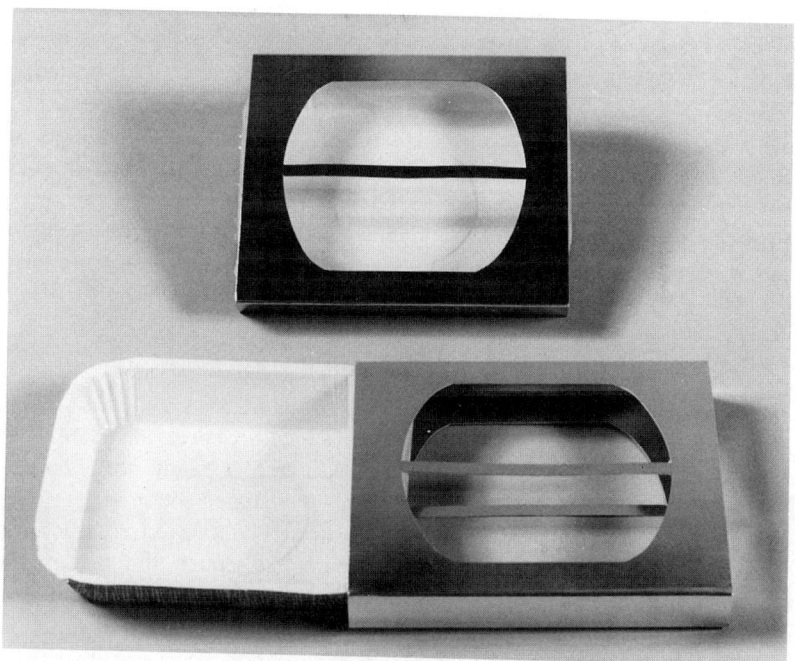

Fig. 1.18 Food tray and shielding device used with U.S. Air Force meal heating unit in Fig. 1.17.

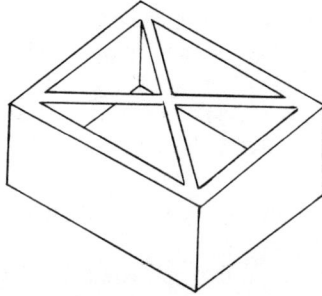

Fig. 1.19 Food package with film covered die-cut lid used specifically for crisp foods in U.S. Air Force meal heating unit.

or these items might be programmed to cook for immediate consumption. A number of these machines in the appropriate locations such as service areas of apartment houses or condominiums could provide a wide variety of cooked, heated, thawed, or raw foods of high quality for the most discriminating customer.

Today, there is renewed activity in microwave vending. One firm under the name Cafe Quick is in the process of completing plans to install a number of vending machine to heat and vend french fries, fried chicken nuggets and pizza. A combination of microwave heating and jet impingement hot air is involved to give a vending time from the frozen state of about one minute. The heating system and vending mechanism were developed by Enersyst Development Center, Dallas, TX (Smith, 1991).

**In-flight feeding**

Harrington (1965), a Trans World Airlines vice president, stated in an article in Air Transport World, "Our goal is to cook a meal a minute from zero degrees. This would give us the flexibility of a completely frozen entree service with waste virtually eliminated because meals would be heated only on order. As long as meals are kept in the frozen state they can be used on subsequent flights." In 1967, the airline industry issued a challenge to the foodservice equipment industry: "Airlines Flying Restaurants Need Meal-A-Minute Heating Equipment," read the headline of an article in Quick Frozen Foods magazine (Anon, 1967). At that time it was predicted that airline feeding would soon reach a millions meals a day and perhaps 2 million meals a day in 10 years.

In an article on airline foodservice in 1974 (Anon, 1974), an airline official was quoted as saying, during a time when a fuel crisis threatened to affect flight schedules, that "money had been wasted in overprovisioning planes to the tune of $8 million per airplane." Some airlines were looking to frozen foods to help reduce waste. All seemed convinced that food would remain a major factor in the future of air travel.

Litton Industries Atherton Division in 1965 designed a microwave oven (Fig. 1.20) for Trans World Airlines and after exhaustive flight testing to verify the absence of electromagnetic interference, a number of these ovens were installed on TWA passenger aircraft. They were used almost exclusively in the First Class section for incidental food heating tasks. Although this application did not proliferate some microwave ovens are being used today on business jet aircraft and Special Air Mission aircraft of the U.S. Air Force including Air Force One, the aircraft used by the President of the United States. These are commercial microwave ovens modified to operate on 400 Hertz power.

The problems of a complete in-flight microwave foodservice system are formidable. Just the energy required to provide hot meals in at least the same time as conventional systems may be too much to warrant its use. On most short flights time does not permit the use of present day batch microwave ovens. Too much handling on the part of already overworked in-flight personnel is required and there is insufficient space for the number of ovens that would be needed.

# HISTORY OF THE MICROWAVE OVEN

Fig. 1.20 Airline microwave oven built for Trans World Airlines by Litton Industries (Courtesy of Litton Systems, Inc., Beverly Hills, CA).

The concept of microwave in-flight foodservice was ill-conceived and did not recognize that the quantities of meals required and the time limitations were too severe for conventional microwave ovens. Competition from relatively inexpensive conventional ovens that heat large numbers of meals at a time was too keen and the amount of handling required in heating one or two meals at a time in microwave ovens placed an inordinate burden on flight personnel. These early ovens designed for aircraft use also were highly sophisticated and costly.

A compact conveyorized microwave system, however, could be designed to handle the meal service on flights of 3 hours duration or longer. A compact version of the Automatic Food Supply system discussed earlier could be designed to fit space available on jumbo jets. Such a concept has been suggested (Anon 1968b). The concept might function as follows: A menu selection panel is located on the back of the seat in front of each passenger. The button beside each item listed is keyed to a totalizer at the galley service panel. After the usual pre-flight preliminaries during which the attendant describes how the system operates, the system is actuated and the passengers are asked to enter their selections by pressing the appropriate buttons. The total number of each meal selected appears automatically on the control panel. These numbers are entered into the meal heating system. The exact numbers of meals requested automatically index out of the freezer storage system, move through the microwave heating system where they are tempered then heated to serving temperature and stored in a controlled temperature holding section. By the time cruising altitude has been reached, the first meals are ready to be placed on trays and served to the passengers. The two stage heating system provides a measure of protection in the event of failure of one or even several of the microwave power modules used to operate the system (Anon, 1968b).

Yet, in spite of the limitations indicated above, a Japanese firm not too long ago (Kawamura and Hiraoka, 1986) announced that specially designed microwave ovens have been installed on Japan Air Lines Boeing 767 aircraft. The 1700 watt oven is said to be able to heat 12 meals in 5 minutes.

**Military feeding**

Some ships of the fleet such as aircraft carriers and submarines use microwave ovens sparingly. On submarines, the oven is used mostly for the convenience of the crew in preparing hot snacks when off duty. A development effort by the U.S. Army Natick Research & Development Laboratories, Natick, Massachusetts initiated in the late 1960s resulted in a prototype all-electric mobile field kitchen and a field bakery that used microwave ovens designed specifically for this application. A related effort was being carried out simultaneously to develop a microwave oven for use in Army field hospitals. These ovens were at the time the most powerful batch ovens ever designed for foodservice.

The SPEED Field feeding program conceived by Snyder (ca 1967), and initiated at the U.S. Army Natick Research and Development Laboratories, SPEED being an acronym for Subsistence Preparation by Electronic Energy Diffusion, progressed through two microwave oven designs (Dungan, 1969; Dungan and Fox, 1969; Decareau, 1971). The first, an all-microwave design that proved to be too limited, was followed by a combination microwave-resistance heating oven design that was unique in that the antenna system that delivered microwave energy to the oven was also the means for providing resistance heat to the oven (Boehm, 1967). Each SPEED microwave oven (Fig. 1.21) was equipped with six power modules, each capable

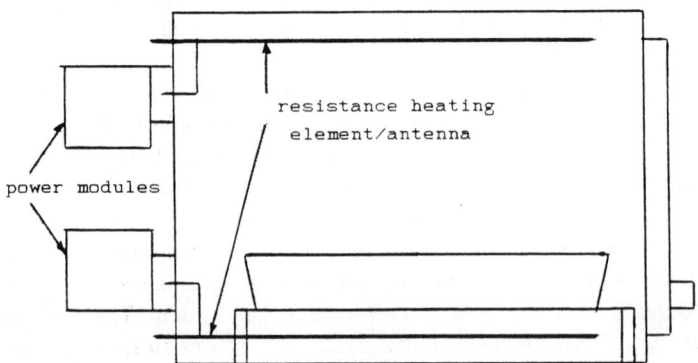

Fig. 1.21 Schematic side view of the thermal/microwave oven built for the U.S. Army's SPEED Field Feeding Kitchen.

of delivering 1.25 kW of microwave power at 2450 MHz to the oven cavity. Three modules supplied microwave power to W-shaped antennas near the oven ceiling and three modules supplied microwave power to similar antennas located near the oven floor. A wire rack shelf was provided to support pans of food above the lower elements. The ovens were equipped with thermostatic means to control oven temperature.

The function of the SPEED kitchen (Fig. 1.22) was to provide foodservice for company-size groups of 200 to 250 men with the microwave ovens to be used for prime cooking of fresh meats and vegetables. This all-electric kitchen was contained in a pod mounted on mobilizer wheels and towed by a suitable military vehicle. The kitchen and bakery were powered by a 60 kW turbine that supplied 400 Hz and 60 Hz power to the kitchen. The lower frequency current permitted the design of compact microwave power modules that made field servicing of the ovens very simple. Because it was desirable to use standard metal roasting and baking pans the lower antennas were operated only as resistance heaters, while both energy means were operative from the upper antennas. Separate controls were provided for each module, while the upper and lower antennas were operated in banks of three for

Fig. 1.22 Partial interior view of the U.S. Army's SPEED Field Feeding Kitchen. The microwave ovens shown were the first generation microwave only ovens. Note the upward opening doors.

resistance heating. Three levels of microwave power were available through controls on the front of the oven. Usable oven space was 37 inches wide by 20 inches deep by 10 inches high. Overall oven size was about 40 by 40 inches by 16 inches high.

The SPEED bakery used the same microwave ovens and would be responsible for supplying bread and pastries for 5000 men per day. Metal loaf pans were used for baking bread. A load of twelve one pound pans of dough that were proofed to near the top of the pan could be baked in about seven minutes to yield excellent, properly crusted bread.

The MUST foodservice system, MUST standing for Medical Unit, Self-contained, Transportable, provided foodservice for MUST field hospital patients and staff. Each pair of MUST microwave ovens served a 60-patient ward complex in heating preplated diets. These ovens used the same antenna feed system as the SPEED ovens except that resistance heating was not provided since only reheating, not cooking, was required. A shelf midway between the ceiling and floor permitted two meals to be heated simultaneously. As in the case of the SPEED ovens, power modules could be removed by loosening four fasteners and sliding the modules out. In the MUST oven the modules were accessible from the front of the oven, while with the SPEED ovens they were accessible through a back panel.

Although neither of these microwave oven programs were adopted by the US Army, much very useful information on combination microwave thermal cooking and baking, and reheating plated meals was generated.

**Railroads**

Microwave ovens were used first by a railroad in the United States when the Pennsylvania Railroad in the early 1950s installed Radarange™ microwave ovens in their Coffee Shop cars for heating refrigerated entrees prepared in their commissary. Some years later the New York Central's New York City to Buffalo run replaced some traditional dining with vending machines and microwave ovens for heating frozen entrees. The entrees were prepared by the 42nd Street restaurant mentioned earlier. Disposable plastic dishes were substituted for china to eliminate warewashing (Anon, 1963b). A concept for using the microwave oven in galleys supporting the dining car was proposed, but not adopted. Later, railways in Europe would use microwave ovens of European design for heating prepared foods in snack bars. Much later microwave forced hot air convection ovens would be adopted by British Rail for use in its dining cars for prime cooking purposes in the gourmet tradition.

**Shipboard**

Other surface transportation systems would adopt the microwave oven to upgrade food service. In the early 1950s Raytheon Company designed a Radarange™ with an oversized oven cavity specifically for the United States steamship line. These ovens were installed on the U.S.S. United States and were in use for a number of years with little or no publicity. Coastal ferries in Scandinavia used microwave ovens and foodservice systems based on microwave ovens, and prepared frozen meals as well as cooking from scratch were proposed for tankers and other ships of the Merchant Marine.

Why the Gingham Kitchen and some of the other microwave oven based concepts did not succeed may never be known. But one argument that has strong support was the inconsistent quality of the products served, due in part to the products, but perhaps to a greater extent to inadequate knowledge of microwave heating techniques. Only in recent years has there been a concerted effort to understand the practical physics of microwave energy heat transfer and the important role of product geometry, power level, conduction heating, dielectric properties and other factors in assuring optimum heating results. Consumers at that time were more demanding and critical of less than the best quality. They insisted on value for their money. We know now that some speed has to be sacrificed if quality is to approach what has long been considered acceptable and normal. A frankfurter in a roll, for example, could only be heated in 10 seconds if the oven power level were close to 2 kW. Such intense heating leaves no time for conduction heating and the result is several very hot spots in the frankfurter, not to mention a dried out roll.

Many of these problems eventually were solved by trial and error and the microwave oven was accepted in many foodservice operations. Hospital foodservice workers learned that the best meal heating results could be obtained by proper arrangement of foods on the plate; i.e., easy to heat foods in the center and hard to heat foods at the edges. Other guidelines that evolved were:

All foods being heated together should be at the same initial temperature.

Frozen foods should not be heated with non-frozen foods.

When heating frozen foods, there should be no signs of partial thawing: food must be solidly frozen when placed in the microwave oven.

Covering foods helps to improve uniformity by taking advantage of the heat from steam generated during heating.

In most cases we know now the scientific reasons behind what was learned the hard way and this will be covered in greater detail in a subsequent chapter.

## THE CONSUMER MICROWAVE OVEN

The first consumer microwave ovens were introduced in the mid-1950s by the Tappan Stove Company under a licensing arrangement with Raytheon Company; by the Hotpoint Company under a similar arrangement in 1956; and later in 1956 the Tappan Stove Company began to private label microwave ovens for several other appliance manufacturers. The Tappan Company deferred to the consumer reaction to the lack of typical browning of foods and added a resistance heating element near the top of the oven (Fig.1.23). The Hotpoint Company chose to offer a cooking center approach with a microwave oven above a conventional oven.

General Electric Company demonstrated the compatability of frozen foods and microwave heating by building a combination freezer-microwave oven that, on demand, would automatically transfer a selected frozen food casserole from the freezer to the microwave oven where it was heated to serving temperature. The XPC-1 (Fig. 1.24), which stood for Experimental Programming Cooker, was a prototype and not intended to be sold. It had space to store 14 different foods any six of which could be heated at one time. Pyrex brand dishes were made especially for this application by the Corning Glass Company. In addition to being microwave transparent

Fig. 1.23 An early consumer microwave oven built by Tappan Stove Company. This oven featured a resistance heating unit mounted near the oven ceiling to brown foods during microwave cooking (Courtesy of White Consolidated Industries, Inc. Cleveland, OH).

Fig. 1.24 The General Electric XPC-1 combination freezer and microwave oven (Courtesy of General Electric Company, Louisville, Kentucky).

these dishes could withstand the thermal shock of being transferred from a freezer to a 400 F oven. The XPC-1 was conceived to be located against a wall adjacent to the dining room so that when the food was ready, as indicated by the sound of a chime, it was transferred mechanically onto a tray and emerged from a small door on the dining room side of the wall ready to be served.

General Electric Company also was the only firm to market a 915 MHz microwave oven. This oven, actually a range called the Versatronic, which combined microwave and resistance heating in a commmon cavity, was introduced around 1963 after years spent in the development of a long-life magnetron. The Versatronic could bake, broil, roast, and also was equipped with surface elements for stove top cooking. Thus it was capable of carrying out the full role of conventional cooking operations as well as microwave and combinations of the two. The oven also was equipped with a turntable for rotating food items through the microwave field to improve microwave cooking uniformity. Actually General Electric had demonstrated electronic cooking to the public as early as 1946. Test marketing of the oven range took place between 1963 and 1965, and active marketing began in 1966. The oven was sold until about 1970, when it became evident that the market belonged to 2450 MHz. This decision was made apparently in spite of the fact that cooking of large food masses such as roasts at 915 MHz gave better results than at 2450 MHz. General Electric later would introduce its own line of 2450 MHz microwave ovens.

The consumer microwave oven market remained relatively quiet until the late 1960s when advances in component technology made it possible to produce a reliable oven at an affordable price. Amana Refrigeration, Inc., which had been acquired by Raytheon Company in 1965, introduced a counter-top microwave oven to the Chicago area in 1967 at a price under $500 (Fig. 1.25). In-store demonstrations were stressed and in-home instruction provided to those who purchased an oven. Amana began to market their ovens nationally in the Spring of 1968. Raytheon microwave expertise and Amana consumer appliance marketing know-how proved to be just the combination needed to stir the market to activity. Significant progress in oven sales was evident with Litton Industries and others entering the market in 1971. The Japanese home oven market was already at the 3500 to 4000 oven per month level and Japanese oven manufacturers were beginning to examine seriously the export market. Sharp, Panasonic, Toshiba, Hitachi, Sanyo microwave ovens eventually competed on the U.S. market under their own name and these manufacturers also competed to build ovens for mass merchandisers such as Sears, Roebuck & Company, Montgomery Ward and others.

Husqvarna in Sweden began its microwave oven development in 1956 and sold its first units in 1958. They introduced their first domestic microwave oven in 1966. Some 2500 were sold in Scandinavia in the 1966 to 1970 time frame. The humidity sensing system used in some Husqvarna ovens was invented by P.O. Risman in 1972. Later patent rights were sold to Matsushita. This was the brain of the so-called Genius oven marketed by Panasonic, Matsushita's microwave oven company. Model 241A

Fig. 1.25 Amana Radarange™ Microwave oven (Courtesy of Raytheon Company, Lexington, MA).

was the first Husqvarna microwave oven to use their humidity control technology and some 500 were produced as a feasibility test. The 206A followed and some 8000 were produced in 1975 to 1977 at which time Husqvarna was purchased by Electrolux and production ceased (Risman, 1991).

### New oven features

Amana Refrigeration, Inc. began the ground work in the beginning of the 1970s that would lead to one of the most innovative microwave oven features perhaps ever to be introduced. The Touchamatic™ Radarange™ microwave oven (Fig. 1.26) was brought out in 1975 and immediately captured the imagination of the public. It had such an effect on the market that other oven manufacturers intitiated crash programs to bring out their own versions. The Touchamatic™ used a microprocessor to program defrosting, cooking, reheating or combinations of these operations. Amana and others began to expand their advertising budgets and the market began to soar.

About 1978 some 10 to 12 per cent of U.S. households were equipped with microwave ovens. At this level of saturation food companies begin to show interest in developing products specifically for an appliance. A number of food companies began to provide directions for microwave oven preparation on their food packages

Fig. 1.26 Amana Touchamatic™ Radarange™ Microwave Oven (Courtesy of Raytheon Company, Lexington, MA).

and still other began to introduce new products. Pillsbury Company, for example, entered the market with microwavable popcorn, pancakes and pizza.

In 1983, 6.1 million microwave ovens were shipped from manufacturers to the marketplace. In 1984 more than 9 million ovens were shipped. In 1985 the volume exceeded 10 million ovens, and in 1986 and 1987 more that 12 million ovens were shipped. Sales have since levelled off in the 10 to 11 million area and it was forecasted that sales would continue at this level at least through 1991. In fact, however due to unforeseen circumstances, primarily a worsening economy, consumer oven sales actually declined significantly in the latter half of 1990 and into 1991. The countertop ovens have been the leader in sales by a large margin whereas the forced convection microwave oven and the microwave range, a conventional oven with added microwave capability, have taken a much smaller share of the market.

Many other interesting features have been added to microwave ovens and only time will tell how lasting they will be. There are ovens that speak to give cooking instructions; ovens that read recipe cards and adjust cooking cycles accordingly; ovens that sense changes in humidity (Risman, 1974) or a burst of food aroma and program the remainder of the cooking cycle; ovens that monitor temperature changes by means of a temperature probe and turn the oven off when a certain pre-set temperature has been reached (Fig. 1.27). Sharp Corporation introduced an oven with a built-in turntable (Fig. 1.28) that has remained a good selling feature for many years. The turntable moves the food being heated or cooked through the microwave field to improve the uniformity of cooking instead of relying on mode stirrers or other means to provide a uniform energy distribution in the oven. Some oven manufacturers added the turntable feature to existing designs that already incorporated mode stirrers. Programmable ovens are prevalent now that this feature has become less costly with volume sales and competition. Recognition devices such as used at supermarket check out counters may be used one day to read coded cooking or heating instructions on food packages. Some prototypes with this feature have been built and demonstrated.

Technology has also reached the stage where more compact ovens have begun to flood the market place. This has resulted in a very substantial decrease in unit cost. The significant increase in sales in 1984 over 1983 was due largely to the introduction of the smaller low powered ovens.

The future of microwave oven sales is exceedingly promising. Soon the household without a microwave oven will be the exception. In 1978 the microwave oven was in fifth place on the list of major appliances in terms of annual sales behind refrigerators, clothes washers, dishwashers and air conditioners. It was predicted (George, 1978) that by 1985 microwave ovens would be Number One in annual sales. That prediction came through on schedule.

Fig. 1.27 Microwave oven with built-in temperature probe (Courtesy of Litton Systems, Inc., Beverly Hills, CA).

Fig. 1.28 Sharp Microwave Oven with built-in turntable (Courtesy of Sharp Corp., Paramus, NJ).

## MICROWAVE OVEN ACCESSORIES

The consumer microwave oven market has spawned another quite lucrative market. Approximately 65% of those who purchase microwave ovens also purchase microwave cookware and other accessories at the same time. Some 19 million units of cookware were sold by 1980 and it was predicted that more than 35 million units would be sold by 1986 at a value in excess of $500 million (Rymer, 1979). Sales of accessories for 1982 were $220 million and the forecast for 1983 was $250 million in sales (Anon,1983).

Some of these accessories are actually appliances that may be characterized as either active or passive. They almost certainly will have a favorable effect on the development of some food products.

An "active" microwave appliance is one that in some way uses microwave energy to generate the heat needed to accomplish a desired effect. Examples of active appliances include: grills, coffee pots, steamers, pizza ovens and browning dishes.

A passive appliance may be described as one that presents the product in the most favorable manner to obtain the best microwave cooking or heating result. Examples include, in addition to a wide selection of dishes, a variety of cooking utensils such as bacon racks, muffin pans, egg pans, roasting racks, tube pans, turntables and some corn poppers.

Microwave cookbooks might also be added to the list of accessories. A recent list numbered more than 100 cookbooks. Hundreds of thousands are being sold and most, if not all, microwave ovens are furnished with a cookbook.

### Microwave corn popper

The passive version of the corn popper is designed to present an aggregate of corn kernels to the energy rather than a single layer of kernels. As the kernels pop the unpopped kernels tend to slide down to the bottom of the conical shaped corn popper to reform the aggregate. They may be made from a number of plastic materials such as polysulfone, filled polyester or polycarbonate.

The active corn popper is designed to concentrate microwave energy at the base of the kernel holder. This is accomplished by creating a high field strength in the region. One manner in which this is accomplished is with a metal ring structure imbedded in a plastic base. Another microwave corn popper uses a cone-shaped bowl set in a resonator that provides a strong microwave field in the vicinity of the kernels (Ishino et al, 1982). The advantage of this over prior art is speed. Prior art as shown in Fig. 1.29 makes use of a holder for the cone-shaped bowl made of a high dielectic constant material, which serves to concentrate the energy in the vicinity of the holder.

### Microwave steamer

This device (Fig. 1.30) is described in the patent of Bowen (1982). The bottom portion is a microwave transparent dish made, for example, of polycarbonate. A

Fig. 1.29 Microwave Corn Popper (Courtesy of Raytheon Company, Lexington, MA).

metal pan with a plurality of holes is supported a short distance above it. A tray for the food to be steamed sits in this pan and a metal cover is placed over the entire assembly. To use the steamer a small amount of water is placed in the bottom plastic dish and the food placed in the tray. When placed in the microwave oven, the water absorbs essentially all of the energy. The steam that is generated passes up through the holes in the metal pan, surrounds and cooks the food on the tray. In short, the food is cooked by steam alone.

## Microwave griddle

This device (Fig. 1.31) is based on ferrite technology and described in patents assigned to Raytheon Company by Freedman and Bowen (1982) and Teich and

Fig. 1.30 Food steamer for use in microwave oven (Courtesy of Raytheon Company, Lexington, MA).

Dudley (1984). This technology will be covered in greater detail in the chapter on browning. Basically, the griddle relies on microwave energy conversion to heat in a ferrite component of this appliance, that in turn transfers the heat by conduction to sear and cook the steaks and chops placed on it.

In practice the steak, for example, is placed on a ribbed metal dish that sits in an insulated plastic base. The hinged cover then is closed down on it. The cover contains a layer of ferrite particles in a plastic matrix in intimate contact with a ribbed metal surface that contacts the upper surface of the steak. The ferrite absorbs microwave energy when exposed in a microwave oven and becomes quite hot (500 to 550 F). This heat is conducted through the ribbed metal surface to the steak to sear and cook it. After a predetermined exposure time the griddle is removed from the oven and the steak turned over for a second cycle to sear and cook the other surface. The ribs of the dish and a trough around its edge drain cooking juices out of the microwave field.

Fig. 1.31 Griddle for use with microwave oven (Courtesy of Raytheon Company, Lexington, MA).

## Cooker/baker utensil

This microwave active appliance patented by Bowen and Martel (1984) also uses ferrite technology to give conventional cooking results in a microwave oven. The utensil is made up of three components: a holder, a metal tray and a cover. The tray and the cover are metallic and shield the contents of the tray. A ferrite layer bonded to the underside of the tray absorbs microwave energy to provide heat which is conducted through the tray to cook food placed thereon. A compartmented version can be used to fry or bake several foods at the same time. There are apertures in the cover to vent steam from the utensil. The base or holder is made of microwave transparent material to support the tray above the oven floor. The base, because it does not absorb microwave energy, also allows the appliance to be handled without danger of burns.

A variation is the microwave pizza oven. It differs in that it has no compartmentation (Fig. 1.32). The energy is transmitted through the base and is absorbed by the ferrite layer which increases in temperature until a temperature, based on its composition, is reached, usually 400 to 500 F. This heat is conducted through the metal tray to cook the pizza. The temperature in the cooking space stabilizes at approximately 400 F. The thermostating temperature is determined by the parameters of the ferrite layer. Because this layer is adjacent to the metal surface where the electric field is essentially zero, substantially all of the coupling is to the magnetic field component and coupling is reduced significantly when the Curie temperature of the ferrite, that temperature at which the ferrite becomes more transparent to microwave energy, is reached.

The microwave transparent base can be made of thermoset polyester with 15–20% by weight of glass or fiberglass. Other materials that can be used include polyphenylene sulfide, polysulphone, Teflon, ceramic or stoneware. The tray may be aluminum or stainless steel, may be Teflon coated and the ferrite layer may be particles of ferrite imbedded in a high temperature silicone based compound.

## SUMMARY

This, then is the microwave oven market and some of the factors that influence it. The food development arena for this appliance has broad boundaries filled with numerous opportunities for creativity. It comes at a time characterized by changing life styles, changes in food preferences, increasing numbers of one and two member households, a growing number of elderly citizens, numerous recreation vehicles, boats, second and third homes, all these providing a market for microwave ovens and products developed specifically for these ovens to make life all the easier and hopefully to improve the quality of life. Microwave ovens offer more certain nutrition by means of precise heating programs moments before consumption of the foods prepared in them. Nutrient losses owing to holding delays and to the discarding of cooking liquid are minimized to the greatest possible extent.

The product developer should be able to approach his target market with confidence that a market of substantial size exists; that is, microwave oven users are actively seeking products specifically designed for the microwave oven. The product developer should have recognized that there are a variety of microwave ovens in use, but that in spite of the many features available they all can be operated as pure microwave ovens. It does not appear to be essential that products for the microwave oven have to be developed to dual-ovenable specifications. The microwave oven alone can support food product lines developed specifically for it. There is ample evidence in the market place that food product developers already have come to this conclusion.

This history would not be complete without some mention of the organization that played a significant role in its development. The idea of a technical society of in-

Fig. 1.32 Pizza oven for use with microwave oven (Courtesy of Raytheon Company, Lexington, MA).

dividuals interested in applications of microwave power was conceived in 1965 at a seminar organized by Drs W. A. Goeffrey Voss and Wayne Tinga at the University of Alberta on the topic: Industrial Microwave Heating. The International Microwave Power Institute (IMPI) was organized as a Canadian Corporation and was off and running in 1966. It was founded for the purpose of furthering the use and understanding of microwave energy for non-information, non-communications applications. IMPI would bring together in annual symposia microwave interests from all corners of the globe. It would catalyze research in all areas of applications of microwave power and result in a vast bibliography of publications not only in the food area but in many unique areas such as plasma etching, ceramic drying, chemical analysis, and cancer therapy to name a few.

IMPI was founded near the end of the lag phase of microwave oven sales when great strides were being made in oven design so that ovens became affordable to a large segment of the population. Home economists and marketing individuals came to look on IMPI as a mecca for the exchange of ideas. The surge in membership from this direction resulted in a skewing of IMPI stress that led in 1979 to the establishment of a separate section — the Cooking Appliance Section (CAS) — devoted to all aspects of microwave cooking, heating, packaging and food product development.

Already the microwave oven was making such an impression on the consumer appliance market that the Association Of Home Appliance Manufacturers (AHAM) would establish a committee to develop standards for the appliance. Membership in the committee came from oven manufacturers, academia, and government. All foreign companies with a vested interest would establish similar committees, conduct studies and participate in international meetings to establish a set of international standards. We now have IEC 705 standards for measuring microwave power that will be used by all microwave oven manufactuers for rating their ovens. This is extremely important for the food product developer who knows now that all ovens are rated in the same manner thus simplifying the preparation of cooking or heating directions on the package.

In addition to IMPI's Journal of Microwave Power, which began publication in 1966, IMPI would sponsor seminars and workshops and make the proceedings available to all interested parties. CAS would introduce its own publication, Microwave World, to cater to the specific needs of their section. The thirst for microwave knowledge has led to many other informative seminars and workshops in recent years both here and abroad on matters such as packaging, microwavable food product development, marketing and others. IMPI also has become the focus for questions on microwave heating and equipment and a voice of authority to respond to occasional misrepresentations in the press about the effects of microwave energy and microwave safety matters.

## REFERENCES

Anonymous (1946). Super high frequency heating for preparation of food. Food Ind. *19*:1699-1700.

Anonymous (1947a). Electronic cooking goes commercial. Electronics *20* (6) 140, 158, 160.

Anonymous (1947b). Frozen meals plus Radarange for speedy service. Quick Frozen Foods *10* (11) 65, 99.

Anonymous (1954). First seminar of microwave cooking. Raytheon Manufacturing Company, Research Division, Waltham, Massachusetts.

Anonymous (1961). Microwave oven makes bid to capture industrial market for mass feeding. Quick Frozen Foods *23* (8) 217-218, 220-222.

Anonymous (1963a). Customers select, heat frozen entrees in speedy microwave ovens at table. Quick Frozen Foods *25* February, 143-146.

Anonymous (1963b). Use of automatic food service by railroad could eliminate conventional dining cars. Quick Frozen Foods *25* October, 159-162.

Anonymous (1966a). Hospital converts to frozen meals by mobile plug-in microwave ovens. Quick Frozen Foods 28.
Anonymous (1966b). The new era. Institutions, Aug.
Anonymous (1967). Mass airline switch to frozen meals inevitable, rate will accelerate. Part I. Quick Frozen Foods, 29 April, 161.
Anonymous (1968a). The microwave oven in the food away from home market. Microwave Energy Appl. Newsl. 1 (2) 7-12.
Anonymous (1968b). Microwaves in airline food service on the ground or in the air? Or not at all? Microwave Energy Appl. Newsl. 1 (4) 9-13.
Anonymous (1974). Airlines chart a clear course for foodservice. Institutions/Volume Feeding, March 1. 37, 39-43.
Boehm, H. (1967). Electronic oven. U.S. Patent 2,320,396.
Bollman, M., Brenner, S., Gordon, L. E. and Lambert, M. E. (1948). Application of electronic cooking to large-scale feeding. J. Amer. Dietet. Assn. 24: 1041-1048.
Bowen, R. F. (1982). Microwave steamer. U.S. Patent 4,317,017.
Bowen, R. F. and Martel, T. J. (1984). Cooker/baker utensil for microwave oven. U.S. Patent 4,486,640.
Brown, B. D. and Doyon, P. R. (1973) An automatic electronic food system. Hospitals, J.A.H.A. 47 (21).
Cease, W. W. (1967). What can be done to increase the productivity of food service personnel. Proc. Soc. for Advancement of Food Service Res.
Constable, R. J. W. (1973). Method and apparatus for cooking. U.S. Patent 3,716,687.
Constable, R. J. W. (1975). Method of cooking food employing both microwave and heat energy. U.S. Patent 3,920,944.
Decareau, R. V. (1971). Electrically powered field kitchens. Cornell Hotel & Rest. Admin. Quart. 12 (1) 73-78.
Dungan, A. L. (1969). The SPEED system after a year of exploratory development. Microwave Energy Appl. Newsl. 2 (1) 7-12.
Dungan, A. L. and Fox, M. A. (1969). The SPEED oven in integrated cooking. J. Microwave Power 4 (2) 44-47.
Ellis, R. F. (1983) $161.4 billion in sales forecast for foodservice industry in 1983. Food Proc. 44 (8) 44-46.
Falk, S. A. (1978). Shielding device for microwave cooking. U.S. Patent 4,122,324.
Freedman, G. and Bowen, R. F. (1982). Ferrite heating apparatus. U.S. Patent 4,362,917.
George, W. W. (1978). The need for voluntary standards. Microwave Energy Appl. Newsl. 11 (4) 3-6.
Harrington, J. E. (1965). TWA goal meal-a-minute by microwaves. Air Trans. World, August.
Hartman, J. (1965). Future hospital kitchens may service satellites. The Modern Hosp., August.
Ishino, K., Miura, T. and Hashimoto, Y. (1982) Popped corn making apparatus used in a microwave oven. U.S. Patent 4,335,291.
Kawamura, K. and Hiraoka, S. (1986). Microwave oven for passenger airplane. Toshiba Review, No. 156, Summer, 37-41.
Moore, N. H. (1966) Microwave energy in the food field. Activities Rept. 18 (2) 163-172.
Park, E. R. (1957). We cook with microwaves. J. Am. Dietet. Assn. 31 (2) 76-80.
Proctor, B. E. and Goldblith, S. A. (1948) Radar energy for rapid cooking and blanching and its effect on vitamin content. Food Technol. 2:95-104.
Proctor, B. E. and Goldblith, S. A. (1951). Electromagnetic radiation fundamentals and their applications in food technology. Adv. Food Res. 3:120-196.
Püschner, H. (1966). Heating with microwaves. Philips Technical Library, Springer-Verlag, New York.
Rejler, M. G. (1970). Pneumatic tube conveyor system with automatic filling and emptying of multiple circulating conveyor containers. U.S. Patent 3,490,717.

Risman, P. O. (1974). Method and device for producing heating of moisture-containing objects. U.S. Patent 3,839,636.
Rymer, J. H. (1979). The impact of the superstar: the microwave oven. Microwave Energy Appl. Newsl. *12* (3) 3–4, 6, 8–10.
Snyder, O. P., Jr. (ca 1967). Subsistence preparation by electronic energy diffusion. Unpublished report, U.S. Army Natick Research Laboratories, Natick, MA 01760.
Smith, D. P. and Harris, H. H. (1974). Factors in design and construction of a device for heating and dispensing food components. Tech Rept. No. 75-41-FL, U.S. Army Natick Research & Development Center, Natick, Massachusetts 01760.
Smith, D.P (1991). Personal Communication, Enersyst Development Center, 2051 Valley View Lane, Dallas, TX 75234.
Sussman, L. (1947). Evaluation of electronic cooking device (Radarange™) for submarines. Bur. Med & Surg., Res. Proj. No. NM 011 016, U.S. Naval Med. Res. Lab., U.S. Naval Submarine Base, New London, Connecticut.
Teich, W. W. and Dudley, K. W. (1984). Microwave heating method and apparatus. U.S. Patent 4,454,403.
Torsiello, J., Reiss, L. and Kirschner, L. R. (1968). Food dispensing machine. U.S. Patent 3,416,429.
Van Gamert, G.A. (1965). All meals in this hospital come frozen. Modern Hosp. *99*, 160–164.
Williams, E. W. (1970). Frozen food forum. Quick Frozen Foods *33* (35) 43.

# CHAPTER TWO

# FUNDAMENTALS OF MICROWAVE HEATING

## WHAT ARE MICROWAVES?

A working acquaintance with microwave heating technology will be helpful to those involved in developing food products for the microwave oven. It is not necessary to be a microwave engineer to understand what happens to food when exposed to microwave energy. A few simple facts will suffice.

Microwaves can be described best as radio waves of very short wavelength. The prefix micro means small. In relation to other radio frequencies, microwaves lie between the television frequencies and infra red. In terms of wavelength, radio waves are measured in kilometers, television frequencies in meters, microwaves in centimeters, and infra red in microns. The relationship between wavelength and frequency is expressed by the simple equation, $\lambda_o = c/f$, where $\lambda_o$ is the wavelength in free space in centimeters, c is the speed of light in centimeters/second and f is the microwave frequency in cycles/second. Thus for a typical microwave oven frequency, 2450 MHz the wavelength ($\lambda_o$) is:

$$\frac{30 \times 10^9 \text{ cm/second}}{2.45 \times 10^9 \text{ cycles/second}} = 12.25 \text{ cm/cycle}$$

The hertz (Hz) is the term used today in referring to the frequency and stands for cycle per second. Thus 2450 MHz is 2450 million cycles per second. This frequency is the only frequency used in microwave ovens manufactured today for commercial foodservice and home use.

Microwaves, like any other wave energy, radiate outward from a source, just as a stone thrown into a quiet pond creates waves that radiate outward in all directions from the spot where the stone broke the surface of the water, These waves carry energy just as the waves that break on the beach, and the amount of energy they carry depends on the amount of energy imparted to them. A vivid example of wave energy is seen in the tremendous destructive force of the waves of a hurricane from energy imparted to the waves by the wind.

The energy of microwaves comes from electrical energy that is converted by a power supply to high voltages that in turn are applied to the microwave power tube or generator to produce power at microwave frequencies. The most common power tube used in microwave ovens is the magnetron and it broadcasts its energy into the oven cavity, as the oven is referred to in the language of microwave oven technology.

## CHARACTERISTICS OF MICROWAVES

Microwaves are reflected, transmitted and absorbed. They behave in the same manner as infra red and light waves. They are reflected from metal surfaces: the oven cavity is basically a metal box in which the waves bounce around. Microwaves are transmitted, that is, they pass through many materials including glass, ceramics, plastics and paper. Some materials are only partially transparent to microwaves; that is, they absorb some energy. When microwaves are absorbed their energy is converted to heat.

### How microwaves produce heat

Microwaves in themselves are not heat. The materials that absorb microwaves convert the energy to heat. In foods, it is the polar molecules that for the most part interact with microwaves to produce heat. Water is the most common polar molecule and is a major component of most foods. The water molecule is called a polar molecule because is has a negative and a positive end (a North and a South pole), and in the presence of a microwave electric field it attempts to line up with the field in much the same manner iron filings line up with the field of a magnet (Fig. 2.1). Since the microwave field is reversing its polarity, millions of times each second, the water molecule, because it is constrained by the nature of the food of which

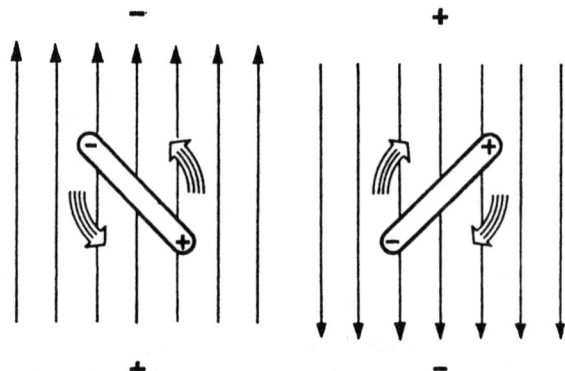

Fig. 2.1 Polar molecules, like water, act like tiny magnets in a microwave field and attempt to line up with the rapidly alternating field.

it is a part, only begins to move in one direction when it must reverse itself and move in the other direction. In doing so, considerable kinetic energy is extracted from the microwave field and heating occurs. The phenomenon is similar to the heating of the human body when exposed to the sun or any other heat source. Energy in the form of infra red rays from the sun, is not heat until it is absorbed by the body and the polar molecules in the surface layers of the body convert it into heat.

Ionic conduction is another important microwave heating mechanism. Ions, being electrically charged, are influenced by microwave fields that cause the ions in solution to flow first in one direction then in the opposite direction as the field is reversed (Fig. 2.2). The effect of ionic conduction can be observed in the microwave heating of salted water in that higher temperatures are found at the surface. Ionic conduction also occurs in cellular fluids when animal or vegetable tissues are exposed to microwave energy.

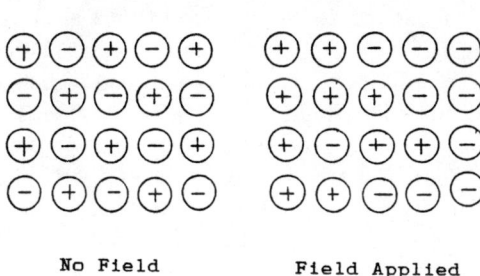

Fig. 2.2 Ions in solution tend to move with the changing electric field.

## Microwave properties of foods

The amount of power that can be absorbed by a substance is expressed by the relationship:

$$P = \sigma E^2 \text{ (watts/cm}^3\text{)} \qquad \text{Eq (2.1)}$$

where  P = the power absorbed in watts/cm$^3$
 $\sigma$ = the equivalent dielectric conductivity
 E = the voltage gradient in volts/cm

The dielectric conductivity,

$$\sigma = 2\pi\epsilon_0\epsilon''f \qquad \text{Eq (2.2)}$$

where f = the frequency of the energy source
$\epsilon_o$ = the dielectric constant of vacuum (8.85 × 10$^{-12}$ farads/m)
$\epsilon''$ = the dielectric loss factor of the substance

Substituting Eq (2.2) into Eq (2.1) gives

$$P = 55.61 \times 10^{-14} f\epsilon''E^2 \qquad \text{Eq (2.3)}$$

where $\epsilon'' = \epsilon' \times \tan \delta$

**Loss factor**

Since the field strength (E) and the frequency are essentially constant for the microwave oven being used, the loss factor ($\epsilon''$) is the only variable. The term "loss" originates from the energy loss that occurs when the electrical energy through a capacitor is cycled on and off. It was realized that not all of the electrical energy was recovered. Some of the energy was given up (or lost) in heating the dielectric of the capacitor. The term today, in microwave heating technology, is representative of a desirable condition. In the context being used, the loss factor represents a property of the material being processed, and a "lossy" material is one that heats well, while a "low loss" material is one that heats poorly and is therefore more transparent to microwave energy. The loss factor is a measurable quantity and a considerable volume of dielectric loss data has been accumulated (Bengtsson and Risman, 1971; Mudgett et al, 1977; Ohlsson et al, 1974; Ohlsson and Bengtsson, 1975; To et al, 1974; Kent, 1987).

The loss factor is the product of two other measurable properties: the dielectric constant ($\epsilon'$) and the loss tangent (tan $\delta$). These values vary with temperature and frequency as shown in Fig. 2.3 (Wang and Goldblith, 1976). They tell us something about the penetration of microwave power. Specifically, they point out that penetration of microwave power decreases dramatically at the lower frequencies as the temperature increases whereas the changes at 2450 MHz are relatively small. Thus the choice of 2450 MHz as heating frequency is fortuitous since it is less sensitive to different loads within a range of moisture content, salt content, and temperatures.

**Penetration**

The penetration depth is that depth in a material at which the microwave power level is 37% of the surface value (or 1/e). The term half-power depth is also used, as in Fig. 2.3, and means that depth in a material at which the power level is one-half that at the surface. The equation for converting dielectric property data into penetration depth (dp) is

# FUNDAMENTALS OF MICROWAVE HEATING

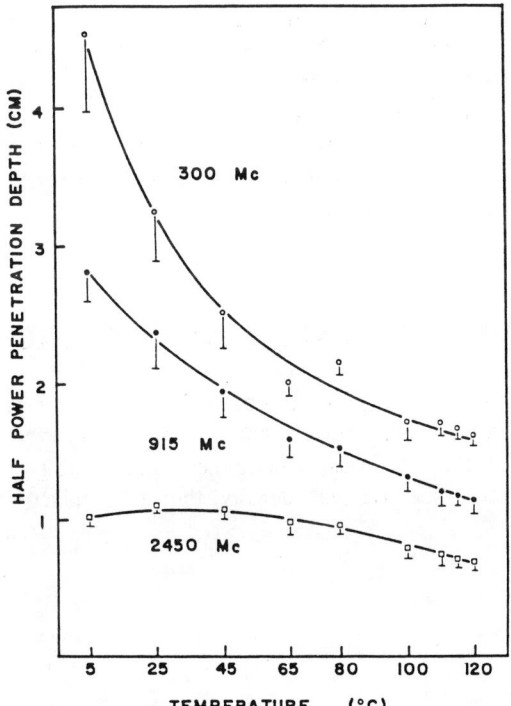

Fig. 2.3 The effect of temperature and frequency on the penetration of microwave power into unsalted meat products (from Wang and Goldblith, 1976).

$$dp = 1/2\alpha \ (\text{for } 1/e)$$

$$dp = 1/2.886\alpha \ (\text{for half-power depth})$$

where $\alpha$ = the attenuation constant

$$\alpha = \frac{2\pi}{\lambda_o} \left[ \frac{\epsilon'}{2} (\sqrt{1 + \tan^2\delta} - 1) \right]^{1/2}$$

A much simpler equation gives results that vary only a few percent from the previous equation.

$$dp = \frac{\lambda_o \sqrt{\epsilon'}}{2\pi\epsilon''} \ \text{for } 1/e$$

Thus the main difference between microwave heating and other heating methods is that microwaves penetrate deeply into food materials and are converted to heat as they penetrate. It is not heating from the inside out as is often said or implied.

The temperature profile shown (Fig. 2.4) illustrates the effect of microwave heating on a large food mass and the effect of time after the energy is no longer being applied. It is clear in this case that conduction heating will always play a part in any microwave heating process where a temperature gradient exists, however, it will exert a greater role with large masses than with small masses. It also clearly shows the effect of evaporative cooling by the lower temperature at the periphery.

## EFFECT OF VARIOUS PHYSICAL FACTORS

In addition to the dielectric properties of foods there are a number of other factors that affect microwave heating performance, perhaps to an even greater extent than by conventional heating methods. Among these factors are geometry (shape), surface to volume ratio, specific heat, density, thermal conductivity and evaporative cooling.

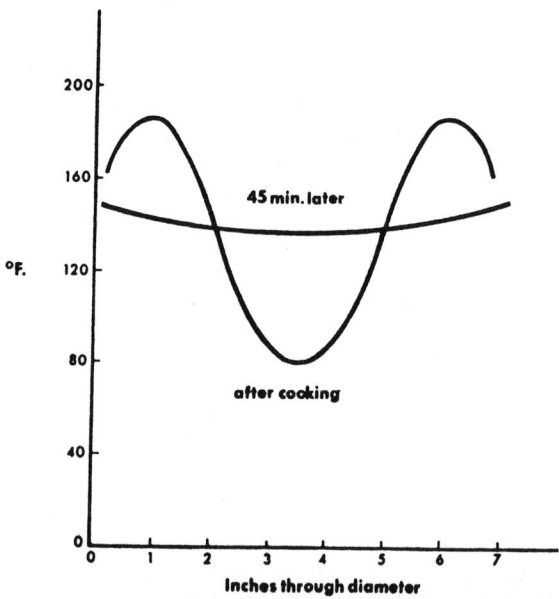

Fig. 2.4 The temperature profile of a 22 lb standing rib beef roast after microwave heating and a prolonged standing period.

## Geometry

The shape of food items is critical to good microwave heating results. The sphere is the ideal shape as energy tends to be focused to give heating at or toward the center of the sphere. Obviously as the diameter is increased it may be impossible for center heating to occur except by conduction. As reported by Ohlsson and Risman (1978), microwave heating at 2450 MHz can be concentrated in the center of spheres with diameters that measure between 20 and 60 mm. Such concentrated heating can be a disadvantage in that conduction heating may not be able to dissipate the temperature gradient and the mass could erupt, as in the heating of an egg in its shell. At lower microwave heating rates, more time is provided for conduction heating to take place. In the example of beef roasting the post microwave heating time could have been reduced if a lower heating rate were used, and the temperature gradients would not have been as great.

Ohlsson and Risman (1978) carried out computer simulation studies as well as actual heating of spheres and cylinders of various diameters. A phantom food mixture with dielectric properties similar to food materials was used in these studies and temperature patterns were obtained by color thermography. The mixture used was essentially that described by Guy (1971):

75% aqueous solution of 0.8% NaCl and 0.2% NaF

17.5% TX-150 water binding agent (Oil Center Research Inc., Lafayette, Louisiana)

7.5% polyethylene powder

To prepare, mix the aqueous solution and polyethylene powder in a plastic beaker. Then mix vigorously while slowly adding the TX-150. Continue mixing until the mixture begins to thicken, then pour into suitable molds. The phantoms are ready for use in 15 to 20 minutes. Molds, preferably, are made in two halves and a thin sheet of plastic film placed between and the assemblage supported in polyurethane blocks. Heating was carried out in a 1300 watt, 2450 MHz microwave oven.

For cylinders, thermography indicated maximum core heating occurred at diameters of 35 mm. At 50 mm, the core and surface were heated, while at 75 mm only surface heating was observed. Calculations support maximum center heating for diameters of 20 to 35 mm, while surface heating was more prominent at 40 to 50 mm and was dominant for larger cylinders.

These results are in qualitative agreement with Copson (1956) for agar cylinders of much larger dimensions. Agar, however, because it contains only a few per cent solids, is more representative of immobilized water than food.

While spherically shaped foods are commonplace among many natural foods e.g., beets, onions, potatoes and some prepared foods (meat balls), the majority of foods are not spherical in shape. The cylinder is the next best shape in terms of heating performance. How this is so can be more clearly demonstrated by the simple example of a rolled meat roast 5 inches in diameter and 10 inches long. The volume of meat within one inch of the surface is 71.5% as shown in Fig. 2.5. This is the region where the majority of the heat is generated as reflected in the penetration of microwave power into meat from Fig. 2.4. It is evident that if heating were continued to a center temperature of 140 F (rare for beef), gross overcooking of the surface would occur. The technique that has evolved for microwave roasting is to cook to a center temperature of about 100 F then allow the heat in the surface region to complete cooking by natural conduction of heat to the center. The dotted line in the figure represents the temperature across the diameter after the roast has been standing for an appropriate period of time. Meat roasts, for example, can be obtained in such a shape and good to excellent cooking results are possible. However, there are many other meat items that do not conform (pork chops, steaks, cutlets, chicken legs, etc). Obviously there is a strong tendency for such foods to be overcooked where they are thinnest. Most microwave ovens today have a roast setting at a somewhat reduced power output.

The donut shape appears to offer a logical solution for heating food products that normally would present a difficult task in heating to the center of the food mass since this shape eliminates the center altogether. A meat loaf in a donut configuration is often recommended in microwave oven cookbooks. Tube pans help to solve this problem in cake baking, and such pans are readily available in microwave transparent plastics for this purpose. Some containers in this shape (Fig. 2.6) have appeared for portions to be heated in vending locations (Watkins, 1983). In this illustration a foil layer around the outside of the bowl provides additional heating control.

For many foods it is the food container or dish that determines the food shape and therefore affects the heating performance. A rectangular shape is not uncommon, but it is easy to see that food in the corners of a rectangular container will be overcooked before the remainder of the food is ready. Overcooking in the corners occurs when using a conventional oven also, but because conduction cooking is much slower the differences are not as pronounced. Where this shape is unavoidable, the corners can be shielded with aluminum foil to reduce the heating rate in these areas and give more satisfactory results. Even when corners are unavoidable they should have generous radii to minimize the overheating effect. Also vertical or near vertical sides are preferable to sloping sides. Excessive sloping gives areas where dehydration can occur.

**Shielding**

Shielding is the use of metal to reflect microwave energy from certain areas of a food item to reduce the heating rate in those areas. Most often the shielding metal is aluminum foil, primarily because of its availability, but also because it can be wrapped around specific parts to be shielded. An interesting use of foil shielding is to place sauces and condiments is appropriate sized foil cups so that as components of frozen dinners they will thaw and not be heated to high temperatures; for exam-

# FUNDAMENTALS OF MICROWAVE HEATING

$V_o$ = 2.5 10 =196.25 cu. in.

$V$ = 1.5  8 = 56.52 cu. in.

$\overline{139.72}$ cu. in

$\dfrac{132.72}{196.25}$ = 71 % of roast is within 1.0 inches of the roast surface

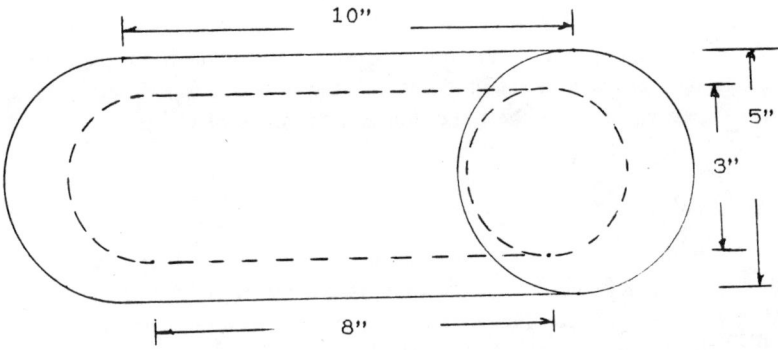

Fig. 2.5 The volume of meat within 1.0 inch of the surface of a cylindrical roast.

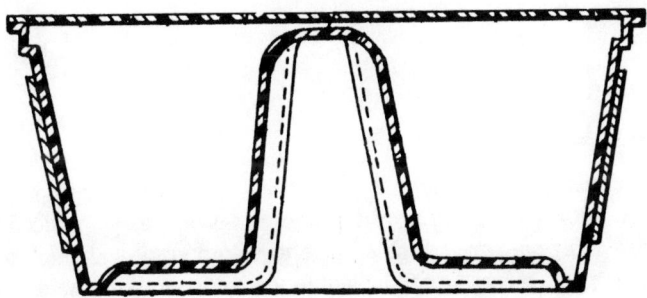

Fig. 2.6 Containers can be shaped to accentuate heating effects (Watkins, 1983).

ple, cranberry sauce as a part of a turkey dinner, or tartar sauce with a fish dinner. Other examples of shielding include wrapping the ends of chicken wings and legs, or the ends of rolled meat roasts (Fig. 2.7) for a part of the cooking cycle. Shielding often must be removed at some point in the cooking cycle otherwise there will be uncooked areas. For example, in meat roasting the shielding on the ends should be removed after about two thirds of the cooking time to allow the ends to cook. Other examples will be discussed in the chapter on packaging.

## Shadowing

Shadowing can be described as a shielding effect by a food product that results in a reduced heating rate in an adjacent product. Thus it is a mutual effect with products shadowing each other. For example, if two meat roasts are placed side by side in contact with each other there will be less cooking in the contact areas (Fig 2.8). Similarly, if several potatoes are grouped together in the oven, the temperature in the contact areas will be much lower than the outside areas. Food items should always be spaced apart for best cooking or heating results.

## Positional relationships

This is another way of saying plating; that is, where to place foods in relation to other foods on a plate to be microwave heated before serving. Some foods heat faster than others and their heating can be slowed by placing them in the center of the plate. Slow heating foods should be placed near the rim of the plate with the thickest areas to the outside. An example is shown in Fig. 2.9. Diced carrots, which heat rapidly because they have so much surface area, are placed in the center of the plate. A pork chop, fat side out, is a slow heating food because of its thickness and density, and is placed at the outside of the plate. Potato, specifically mashed, creamed or any other form of potato that gives a cohesive mass, likewise is placed at the outside for the same reason. Diced potato, or potato in any other loosely held together form with a high surface to volume ratio would be treated as a fast heating item.

## Surface to volume ratio

As with conventional heating, the greater the surface area, the faster that cooking occurs. Thus potatoes and carrots can be cooked faster if they are diced. Vegetables such as corn and peas have a high surface to volume ratio and cook rapidly. This same condition also means these kinds of foods will cool more rapidly. This should be kept in mind when developing products, and means to reduce the cooling rate provided where possible; for example, by using insulated containers or preheated serving dishes.

Fig. 2.7 Shielding of the ends of a roast with aluminum foil reduces the heating in this area so that the remainder of the roast can be cooked without overcooking the ends.

## Specific heat

The specific heat of a material is the ratio of its thermal capacity to that of water. The specific heat of a food is closely related to its moisture content. A convenient mathematical relationship known as Siebel's formula is useful in estimating the specific heat when the approximate moisture content is known

Above freezing:

$$\text{Specific Heat} = 0.2 + 0.008a$$

Below freezing

$$\text{Specific Heat} = 0.2 + 0.003a$$

Where a = the per cent moisture as a whole number.

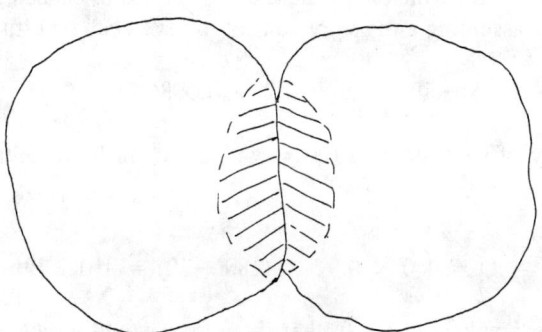

Fig. 2.8 Foods placed against each other will be undercooked where they are in contact.

58 MICROWAVE FOODS: NEW PRODUCT DEVELOPMENT

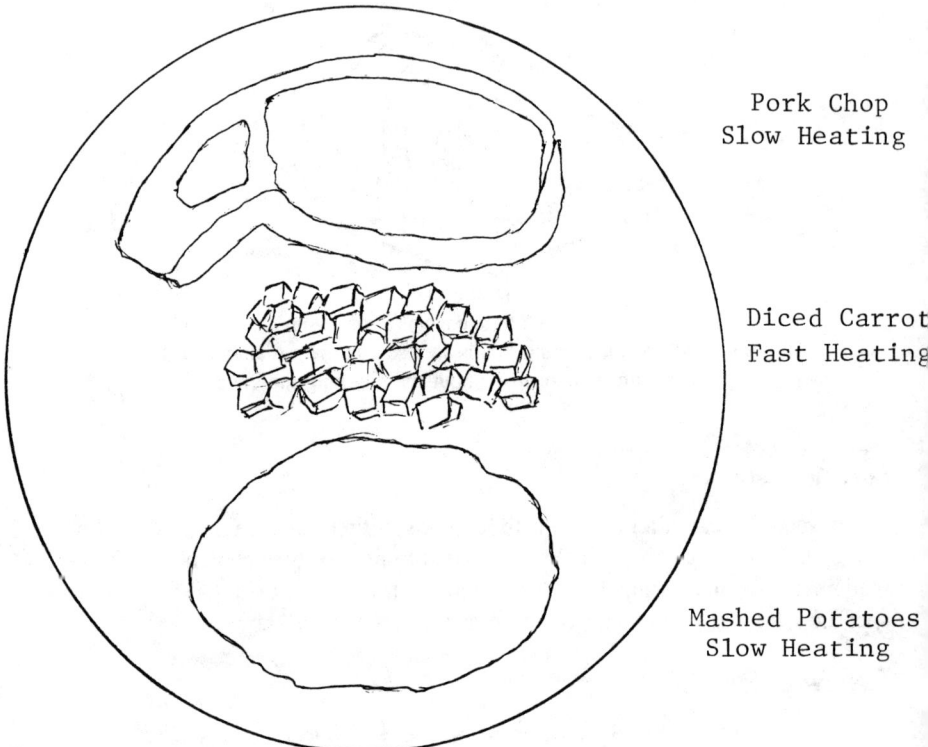

Fig. 2.9 When plating foods for microwave heating it must be recognized that some foods heat faster than others.

The specific heat enters into the formula for calculating the heat required, in British thermal units or calories, to accomplish a specific heating task. For example, the energy required to raise the temperature of one pound of canned green beans from 70 F to 180 F, assuming a moisture content of 90%, can be estimated as follows:

$$\text{Specific Heat} = 0.2 + 0.008(90) = 0.92$$

The quantity of heat required (Q) = wt (lb) × sp. ht × delta T(F)

Thus

$$Q = 1.0 \times 0.92 \times (180 - 70) = 101.2 \text{ Btu}$$

To carry the example one step further, how long would it take to heat one pound of canned green beans in a 600 watt microwave oven assuming 100% conversion of microwave energy to heat?

# FUNDAMENTALS OF MICROWAVE HEATING

$$1 \text{ kWh} = 3413 \text{ Btu/h or } 57 \text{ Btu/min}$$

Therefore 600 watts = 3413 × 0.6 = 2047.8 Btu/h or 34 Btu/min

$$\text{Heat required/Energy input} = \text{time}$$
$$101.2 \text{ Btu}/34 \text{ Btu/min} = 2.95 \text{ minutes}$$

## Microwave heating efficiency

This is an appropriate place to discuss heating efficiency, particularly since in many cases relatively small quantities of foods will be heated at a time. Fortunately a good deal of data has been developed on this aspect of microwave heating.

The procedure for measuring the power output of a microwave oven is to heat a measured quantity of cold water for a specific period of time, noting the temperature change. In the procedure that is widely used in the microwave oven industry, one liter of water in a Pyrex glass beaker is heated for two minutes. The initial temperature of the water should not exceed 20 C in order to avoid the situation where heating for two minutes gives a final temperature where evaporation is occurring. Evaporation consumes much energy and would skew the results unfavorably. Using the metric system a conversion factor is necessary to obtain the results in watts

$$\text{Power (watts)} = \frac{1000 \text{ gm} \times 1.0 \times \text{delta T(C)} \times 4.189}{120 \text{ seconds}}$$

Combining all the known values gives a constant of 34.9 (round to 35). Thus

$$\text{power} = \text{delta T(C)} \times 35$$

A temperature change of 20 C therefore would represent a powerlevel of 20 × 35 or 700 watts.

When the power measurement is carried out with lesser quantities of water, the heating time is reduced proportionately. The power levels are shown to decrease (Fig. 2.10) as the efficiency with smaller loads decreases and heating time may need to be increased to achieve the desirable temperature.

## Density

Usually there is a clear relationship between density and moisture content. Thus bread, which has a low moisture content (about 35%) is not as dense as beef (about 65% moisture). Referring back to the equation for determining heat required it can be seen that one pound of bread requires about two-thirds as much energy as one pound of beef to heat to the same temperature. Thermal conductivity also comes

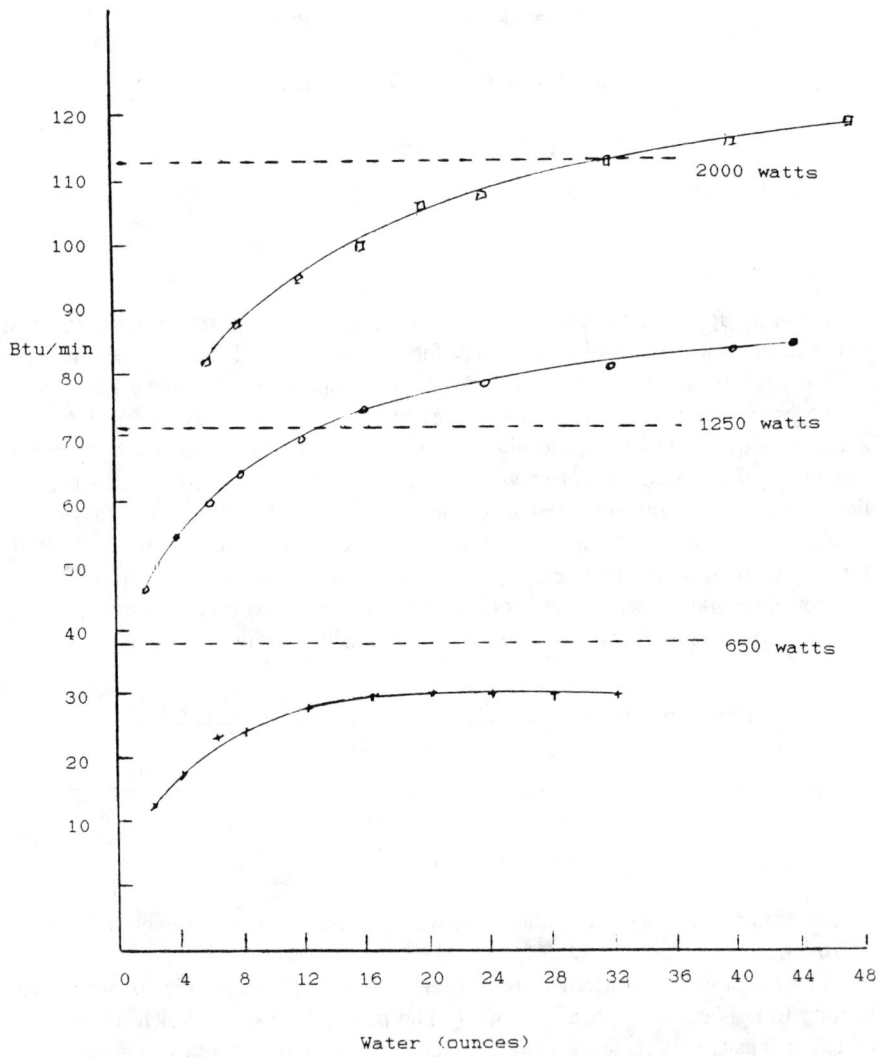

Fig. 2.10 The effect of load size on microwave power absorption.

into play in that the dimensions involved could mean substantial overheating of the surface of beef because of its density, while the more open structure of bread would not rely on thermal conductivity to the same extent. From a practical point of view, for example, heating an assembled hamburger sandwich, the bread component would receive substantially more heating than the hamburger pattie. It would be overheated driving out moisture, thereby adversely affecting the quality of the bread component, while the hamburger would be underheated. This problem has a solution as will be shown later in the chapter on packaging.

## Thermal conductivity

Thermal conductivity is a measure of a material's ability to transfer heat in response to a temperature difference. Conduction heat transfer depends on a temperature difference. Even in microwave cooking it plays an important role in spite of the penetrating nature of microwave energy. Without heat conduction there would occur unacceptable temperature differences in most microwave heated or cooked foods. Such differences can be minimized by reducing the rate of microwave heating.

Thermal conductivity also plays a very important role in the thawing and heating of frozen foods. Frozen foods are better conductors of heat than non-frozen foods, which explains why frozen foods thaw very slowly by conduction heating methods, and why foods can be frozen by conduction methods much faster than they can be thawed. The difference in thermal conductivity is about 3 to 1 in favor of frozen foods.

Microwave energy makes it possible to raise the temperature throughout a frozen food very rapidly up to a point. This is the point at which the specific heat begins to increase rapidly so that external layers of the food become completely thawed and behave as a resistance to further deep microwave heating. The transition point for beef is about 26 F, and is much lower for foods that have lower freezing points such as juice concentrates and ice cream.

Thus microwave energy can be used for complete thawing without unacceptable temperature gradients only by reducing the energy application rate after the transition temperature has been reached. This is not always practical and detracts from the major advantage of microwaves; i.e., speed. However, the two-step process should be used with some adjustments for most microwave heating applications; for example, where permissible, the product should be broken up after it has been tempered to expose more surface area to accelerate the final thawing step. A cogent example of how this can be done with a prepared, frozen boil-in-the-bag food product is illustrated in Fig. 2.11. A typical pouch of frozen food is heated for about 1.5 minutes in a 600 watt oven. This should be done without puncturing the film. The product is usually soft enough after this heating time to massage and to shake down in the pouch so that it can stand upright in a serving dish for final heating. At this time the bag should be punctured to release steam during heating to completion.

## Starting and final temperature

Heating time depends not only on the power level but also on the temperature gap that must be bridged to accomplish the desired result. Thus frozen foods take longer than refrigerated foods, which in turn take longer than shelf stable food heated from room temperature assuming all were heated to the same final temperature. An additional quantity of heat is required to heat from the frozen state (the heat of fusion) which is 144 Btu/lb of ice, and proportionately less for foods depending on the moisture content.

## Evaporative cooling

A phenomenon that is much more evident when cooking with microwave energy

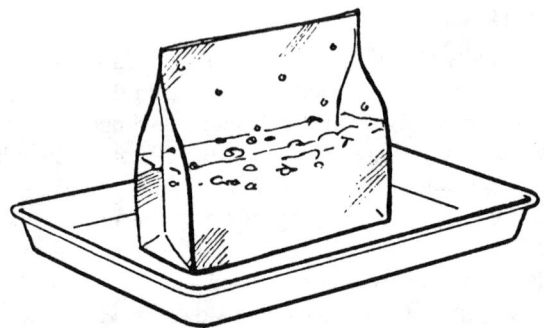

Fig. 2.11 One way to microwave heat food packaged in a boilable bag.

is evaporative cooling. This effect is responsible for the lower surface temperature of some foods cooked in microwave ovens and therefore responsible for the myth that microwaves cook from the inside out. It is not noticeable in a conventional oven because of heat radiation from the oven walls to the surface of foods. Proof of this effect can be found in the fact that when foods are blanched with microwave energy to inactivate enzymes, those enzymes near the surface may not be completely inactivated unless the foods are heated in a closed package to permit the steam generated to raise the surface temperature sufficiently to inactivate them; or microwaves and steam are combined in the blanching process.

## EFFECT OF VARIOUS OVEN RELATED FACTORS

There are a number of other factors affecting microwave heating performance of which the product developer needs to be aware, but over which there is little control. Schiffmann (1991) addressed this subject at a recent symposium and cited, among others: output wattage, electric field distribution, presence or absence of a turntable, type and location of the energy feed system, cavity volume, cavity material, power supply, glass or metal shelf, oven floor material, time base for pulsing power, line voltage, age and condition of the magnetron power tube and cold vs warm start of the magnetron.

### Oven power

As shown earlier, oven power is sensitive to load size as well as to how it is measured. Schiffmann (1991) presented data that clearly demonstrated that, for example, an oven rated by the IEC 705 standard can give a different heating result than another oven rated by the traditional method; i.e., heating a liter of tap water for two minutes. The first oven gave a reading of 800 watts while the second rated only 600 watts. When they were used in popping a bag of popcorn, the first oven required 4.5 minutes while the second only 3.5 minutes. Popping volume was also significantly better with the second oven.

### Electric field distribution

The uniformity of the electric field may vary significantly from oven to oven and even among ovens from the same production run. Various methods for characterizing microwave oven field patterns have been used over the past 40 years or so. Stanford (1990) described a simple method in which heat sensitive paper and ordinary corn meal was used. The heat sensitive paper is placed on a supporting sheet of corrugated paperboard with the dimensions of the oven floor. Corn meal is sprinkled evenly over the entire surface of the paper. After heating for a predetermined period of time the paper is removed and the corn meal brushed off revealing a dark pattern resembling an ink blot. The non-colored areas reflect cold spots. For best heating results food products should be placed in the dark zone locations in the oven.

### Turntable

Although ovens with built-in turntables provide somewhat more even heating of food products than ovens without them it is still possible to have hot and cold zones. Metal turntables may give poor heating results with thin profile foods or foods placed on susceptors because of the low field strength at the metal surface.

### Feed system

In most microwave ovens the energy is transferred from the magnetron to the oven via a waveguide. This metal conduit usually opens through the top of the oven cavity where the energy is intercepted and reflected off a rotating device called a mode stirrer. The electric field distribution in an oven without a mode stirrer is incredibly uneven. Some ovens have feed ports on opposite sides of the oven and some are fed from the top and bottom of the oven. Each type has its own peculiar electric field pattern.

### Cavity volume

Microwave oven cavities come in a wide variety of sizes from less than 0.5 cu ft to over 1.5 cu. ft. The larger cavity usually will have a better electric field distribution than a small cavity because it can sustain many more nodes.

### Cavity material

Microwave ovens may be fabricated from stainless steel, acrylic coated cold rolled steel or aluminum. The electrical conductivity of these materials differ and the more conductive the material the more power that is available to heat the load.

### Power supply

The output of the power supply is affected by the input voltage, and can vary substantially. For controlled laboratory work a voltage regulated power supply is desirable.

### Oven floor material

The oven floor, which defines the lower limit of usable oven volume, may be glass, low absorbency ceramic, or bare metal. The function of a glass or ceramic floor is to hold the food load a short distance above the metal floor so that energy can be reflected from the floor up into the food. Usually these materials are highly transparent to microwave energy, however, and even tempered glass trays used in some ovens to support the food will absorb some energy.

### Time base

Most microwave ovens provide variable power by pulsing the microwave energy. The time base on which pulsing is carried out may vary from a few seconds to 30 seconds or even more. Thus 50 per cent power could be one second on and one second off or less or as much as 30 seconds on and 30 seconds off. A large time base makes it unrealistic to use any power level but full power.

### Line voltage

Voltage can vary significantly throughout the country. In summer, brownouts sometimes occur when a large percentage of the population turn on their air conditioners. The low voltage available at such times means that microwave oven power will be low making it impossible to do some things in some ovens such as pop corn.

### Age and condition of the magnetron

Usually the magnetron has a very long useful life, and indeed it may be less costly to purchase a new oven than to have a new magnetron installed.

### Cold start versus warm start

Power output after several minutes of use usually stabilizes at a power level somewhat lower than from a cold start.

Because there are so many factors that can affect microwave heating performance the product developer must test his products in as wide a variety of ovens as possible and provide general heating directions. The oven user must always realize that best results can only be assured by checking the temperature of food products after heating directions have been followed and adding increments of energy to complete the heating process if necessary. Some oven manufacturers have recognized the limitations of their ovens and provided a separate control to add energy in one minute, more or less, increments.

# REFERENCES

Bengtsson, N.E. and Risman, P.O. (1971). Dielectric properties of foods at 3 GHz as determined by a cavity perturbation technique. II. Measurements on food materials. J. Microwave Power 6:107–123.
Copson, D.A. (1956). Microwave energy in food procedures. IRE Trans PGME-4:27–35.
Guy, A.W. (1971). Analysis of electromagnetic fields induced in biological tissues by thermographic studies on equivalent phantom models. IEEE Trans. MTT 19(2), 214.
Kent, M. (1987). Electrical and dielectric properties of food materials. Science & Technology Publishers, Ltd. London.
Mudgett, R.E., Goldblith, S.A., Wang, D.I.C., and Westphal, W.B. (1977). Prediction of dielectric properties in solid foods of high moisture content at ultrahigh and microwave frequencies. J. Food Proc. Pres. 1:119–151.
Ohlsson, T., Bengtsson, N.E., and Risman, P.O. (1974). The frequency and temperature dependence of dielectric food data as determined by a cavity perturbation technique. J. Microwave Power 9:129–145.
Ohlsson, T. and Bengtsson, N.E. (1975). Dielectric food data for microwave sterilization processing. J. Microwave Power 10:93–108.
Ohlsson, T. and Risman, P.O. (1978) Temperature distribution of microwave heating-spheres and cylinders. J. Microwave Power 13:303–310.
Schiffmann, R.F. (1991). Oven considerations for the food product developer. Presented at 26th Microwave Power Symposium, Buffalo, NY, 5–7 August.
Stanford, M.A. (1990). Oven characterization and implications for food safety in product development. Microwave World 11 (3).
To, E.C.H., Mudgett, R.E., Wang, D.I.C., and Goldblith, S.A. (1974). Dielectric properties of food materials. J. Microwave Power 9:303–316.
Wang, D.I.C. and Goldblith, S.A. (1976). Dielectric properties of foods. Tech Rpt. TR-76-27-FEL, U.S. Army Natick Res. & Dev. Comm., Natick, MA 01760.
Watkins, J.D. (1983). Microwave food heating container. U.S. Patent 4,416,906.

# CHAPTER THREE

# THE MICROWAVE OVEN

## INTRODUCTION

The food product developer should have an appreciation of the appliance for which the food products are being developed. It is particularly important to be aware of some of the novel features that have come along and the changes that can be expected in the future that could impact on product development opportunities and decisions.

The basic components of the microwave oven (Fig. 3.1) are: (1) the oven cavity; that is the usable cooking space; (2) the door that provides access to the cooking space; (3) the microwave generator, called a magnetron, that generates the energy that does the cooking; (4) the waveguide, the rectangular duct that carries the microwave energy from the magnetron to the oven cavity; (5) the mode stirrer, often a fan-like device that acts as a reflector of microwaves to distribute the microwaves more uniformly throughout the oven cavity; (6) the power supply that converts alternating current from the wall socket to the proper direct current voltage needed to operate the magnetron; and the controls that permit adjustment of microwave power and cooking time.

**The oven cavity**

The oven cavity or cooking space is usually a rectangular metal box, although some experimental ovens were hemispherical and it is not inconceivable that such a design might return one day. The cavity may be made of stainless steel, aluminum or acrylic coated cold rolled steel. Most ovens are designed to be resonant; that is, microwaves will be reflected from all sides to give a pattern of nodes and anti-nodes (or hot and cold spots from a heating point of view). The oven dimensions can vary considerably and the volume can range from about 0.5 cubic feet to 1.5 cubic feet (more or less) for consumer ovens. Actual usable oven space may be considerably less than the oven volume. This is a more important consideration from a cooking view point than from the point of view of reheating precooked foods. The latter may require much less space. Some commercial ovens are substantially larger. There is usually a shelf, a ceramic floor or a tempered glass tray in each consumer oven. This shelf, floor or tray serves to position the food load above the metal floor so that microwave energy can be reflected from the floor up into the food being heated

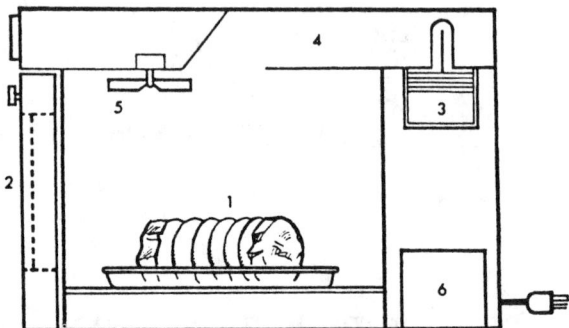

Fig. 3.1 Basic components of the microwave oven: 1, oven cavity; 2, door that accesses cooking space; 3, magnetron (microwave generator); 4, waveguide; 5, mode stirrer; 6, power supply.

as well as from all other directions. In some ovens there also may be a wire rack shelf located well above the floor to permit cooking foods on two levels simultaneously. The glass tray or ceramic shelf serves also as a protective load against empty oven operation.

**Oven heating pattern.** In the design of microwave ovens, engineers often use the term illumination when speaking of the distribution of microwave energy in the oven cavity. The problem is analogous to lighting a room. There are bright, well lit areas and there are areas in a room that are in shadow. These differences are visually apparent and we can make adjustments to improve the illumination of the room. In the microwave oven we have to make the differences in illumination visible in order to be able to improve the distribution of microwave energy. So far the techniques for accomplishing this have been only partially successful.

The simplest and the oldest method for determining the pattern of energy distribution in a microwave oven is to place an ordered array of small containers, each with the same quantity of water, on the oven floor, and to measure the temperature rise in each after a brief heating cycle. A standard procedure described by the International Electrotechnical Commission (IEC, 1988) utilizes a rectangular tank with 25 equal sized wells (Fig. 3.2). Five hundred milliliters of cold water is poured into the tank that has holes in the side walls of each well so that the water level is uniform throughout the tank. After a period of heating in a microwave oven the temperature of each well is recorded to provide an indication of the microwave field distribution for the oven. An example of the kind of results that are obtainable is shown in Fig. 3.3 in which four similar tanks were used on two levels separated by a sheet of aluminum foil to measure the field distribution in a commercial oven with both top and bottom microwave input.

Other methods that have been used include heating pancake batter spread on a piece of glass the size of the oven floor; blotting paper moistened with a solution

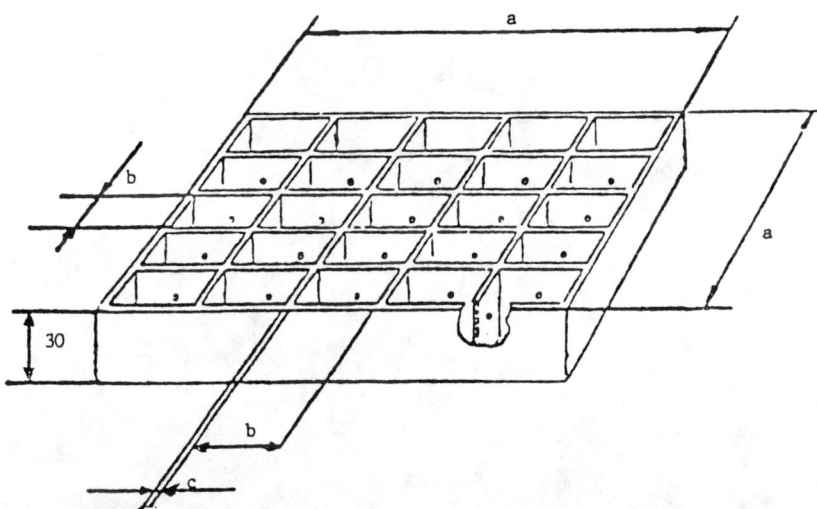

Fig. 3.2 Plastic tank used for assessing the microwave field distribution in a microwave oven as prescribed in IEC 705, 2nd Edition. Dimensions for a 500 cc water load are: a, 168 mm; b, 31.2 mm; and c, 2 mm.

of cobalt chloride, which turns from blue to red as it dries thus leaving a picture of the oven pattern; photocopy paper that turns black where it has been heated; and a variety of food materials, measuring the degree of cooking by temperature and other measures of cooking performance, such as texture, firmness, color, etc. Copson (1975) described a series of cooking operations with means for scoring the results to arrive at a Figure of Merit for an oven. Thus it was possible to evaluate, at the time, a number of prototype microwave ovens and compare them on the basis of their Figure of Merit. A variation on this procedure was developed by Amana Refrigeration Inc. for evaluating their microwave ovens (Peterson and Foerstner, 1971). Other procedures were reported by Kumpfer (1972), Ringle and David (1975), Bobeng and David (1975), Wilhelm and Satterlee (1971), and Berntsen and David (1975). Among more recent procedures can be cited the Stanford (1990) technique, which uses a layer of corn meal sprinkled over thermally sensitive paper to produce an ink blot like picture of the oven heating pattern; and the Swedish Food Institute (Thorsell, 1991) method. The latter is based on the use of a carrageenan gel with food-like properties that can be molded in various shapes to represent various food configurations.

The international standards (IEC 705) of which the second edition was published in 1988 was prepared by Sub-committee 59H: Microwave appliances, of Technical Committee No. 59: Performance of household electrical appliances. This standard involves input from France, Italy, Norway, Sweden, Switzerland, Germany, the United Kingdom, Japan and the United States. This might be called a living document since it is being continually updated.

| | | | | | | | |
|---|---|---|---|---|---|---|---|
| 100 | 100 | 103 | 107 | 105 | 105 | 107 | 130 |
| 105 | 95  | 95  | 95  | 100 | 95  | 95  | 100 |
| 100 | 90  | 90  | 95  | 90  | 90  | 90  | 95  |
| 100 | 90  | 90  | 95  | 90  | 89  | 90  | 95  |
| 105 | 90  | 93  | 97  | 92  | 92  | 95  | 100 |

BOTTOM

| | | | | | | | |
|---|---|---|---|---|---|---|---|
| 115 | 109 | 109 | 115 | 115 | 105 | 107 | 110 |
| 100 | 95  | 95  | 100 | 97  | 92  | 100 | 105 |
| 100 | 95  | 95  | 100 | 95  | 90  | 95  | 100 |
| 95  | 997 | 95  | 100 | 95  | 90  | 90  | 100 |
| 125 | 110 | 103 | 110 | 107 | 100 | 105 | 110 |

TOP

Fig. 3.3 Typical temperature distribution obtained using a variation of the IEC 705 tank. Four tanks with 20 wells each were stacked two on two with a sheet of foil between the two levels to measure the field distribtuion in a top and bottom fed commercial size oven.

## The oven door

The door is one of the most important features of a microwave oven. It not only provides access to the oven space, but when closed must prevent emission of microwave energy from the oven during operation. This is a federal requirement (Anon, 1970b) and strict standards for microwave emission (leakage) have been established. More on this later.

Considerable ingenuity has gone into the door design so that the microwave emission level of microwave ovens is well below the federal standards. The law also requires that microwave oven doors be equipped with two interlocks and an interlock monitor that must function for the oven to operate when the door is closed and yet interrupt the generation of microwave power at the first slight opening of the door. It is a tribute to microwave engineering expertise that today's ovens are able to function with emission levels well below the established emission standards, not just ex-factory, but over the useful life of the oven.

There are two basic types of door seal techniques in use: the choke seal and the contact seal. The former uses a quarter wave slot, or "choke" in microwave language, around the perimeter of the door. The slot is usually filled with a dielectric such as polyethylene to prevent soil from accumulating in the slot. Oven doors using this technique do not require positive contact with the oven face to be effective. Doors with contact seals depend on good metal to metal contact over the entire seal area that when in good condition approaches zero microwave energy emission. Such door seals must be kept clean of food spills and spatters that could provide a pathway for microwave emission. The choke seal is dominant in today's microwave ovens. For greater detail see Decareau and Peterson (1986).

As a measure of the quality of microwave oven door design, the Department of Health & Human Services discontinued their field test program of measuring leakage of microwave ovens because the ovens rarely leaked above the standards unless they were abused or serviced improperly (Barron, 1991).

**The magnetron**

The magnetron is a vacuum tube that uses a built-in magnetic field to affect the flow of electrons from the cathode of the tube to the multicavity anode in such a manner that the electric field is made to oscillate millions of times each second. The microwaves and the energy they carry are then broadcast into the oven cavity. The magnetron is a very reliable device with a normal operating life in excess of ten years. It has been designed to cope with such stress conditions as operating into an empty oven without being damaged or otherwise adversely affected. An example of a magnetron is shown in Fig. 3.4. Current magnetrons weigh less than one pound.

**The waveguide**

Although in some early microwave ovens the magnetrons were directly inserted into the oven cavity, usually through the ceiling, this practice was eventually discontinued for a number of reasons. Being exposed in this way the magnetron became soiled with food spatters and fat aerosols. It was also subject to damage from absorption of reflected microwave energy and could be accidentally broken. Early magnetrons had a tempered glass envelope that could soften if overheated, or shatter if accidentally struck. Perhaps most important, it was difficult to properly and accurately match the magnetron to the oven (tune the oven). Direct insertion was replaced with a waveguide feed.

The waveguide is a section of rectangular metal duct with an open end fixed over a corresponding opening in the ceiling or wall of the oven cavity; and a closed, or shorted end, from which the magnetron power tube is located a specific distance. Microwave energy from the magnetron flows down the waveguide and into the oven. The length of the waveguide is critical to the efficient delivery of microwave energy into the oven. In some ovens the waveguide is designed so that the microwave energy can be delivered into the oven from both side walls or into the top and bottom of the oven simultaneously. An example of the latter is shown in Fig. 3.5.

Fig. 3.4 Microwave oven magnetron.

**The mode stirrer**

Microwaves entering the oven cavity from the waveguide fixed to the top of the oven are intercepted by the mode stirrer, a fan-like device located just below the oven ceiling, that in rotating a few revolutions per minute causes multiple reflections of the energy and in doing so minimizes the number of hot and cold spots that would normally be present. It may be motor driven or propelled by an airstream off the magnetron cooling system. The mode stirrer may take on a number of forms. A pair of them are shown in Fig. 3.6. There are many variations. The mode stirrer is usually protected by a plastic cover to prevent it from being soiled by foodspatters.

It is also possible to improve the heating uniformity by moving the food itself through the microwave field and some ovens have been equipped with turntables to accomplish this. Some ovens have both a mode stirrer and a turntable, and there are portable turntables that can be purchased to use in microwave ovens.

**The power supply**

The power supply is made up of transformers, capacitors, diodes and the wiring circuit that converts the alternating current from the wall socket into high voltage direct current to operate the magnetron. Although this component accounts for a high percentage of the weight of the oven, it is remarkably compact, reliable and efficient.

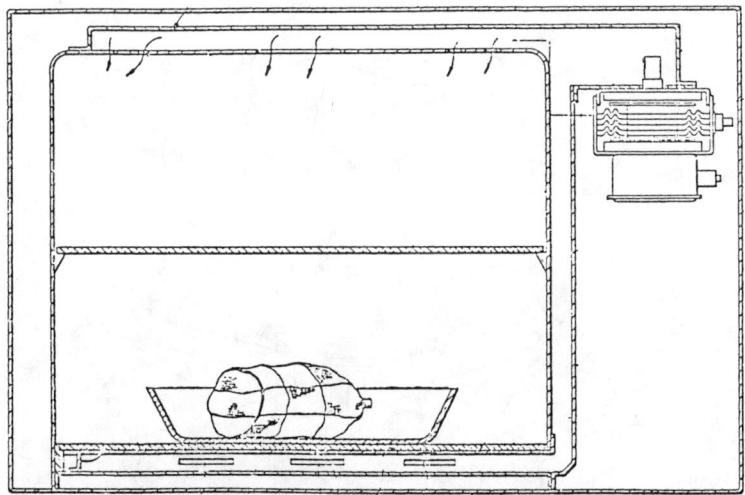

Fig. 3.5 Leaky wall waveguide feed to the top and bottom of the oven from a single magnetron (Staats, 1982)

In 1990, Sharp Corporation introduced a microwave oven with a solid state power supply. This design eliminates a heavy transformer and reduces the overall weight of the oven about 7.5 lb. It is highly likely that this design change will be adopted by other oven manufacturers. Of greater importance to the product developer is the significant improvement in power control provided by this innovation. Instead of pulsing power to operate at reduced average power, direct reduction of power without pulsing is possible down to about 50 per cent power.

**The controls**

Controls on early ovens, before about 1956, were very basic and included a dial timer, a cook button and indicating lights. Today's consumer microwave ovens are equipped with a variety of controls, including dial timers, digital timers and solid state touch pad timers, as well as microprocessor controls, memory features to permit the oven to be programmed and variable power. Commercial microwave ovens used in restaurants and other foodservice operations are somewhat less sophisticated, but do have most of the above features as options. In addition, commercial microwave ovens may have a number of preset push buttons or touch pads for use in vending locations.

There are also a number of other unique oven control features that are worth mentioning.

**Temperature probes.** Some ovens have a wall socket inside the cavity into which a temperature probe on a flexible cable may be fixed. The probe can be inserted

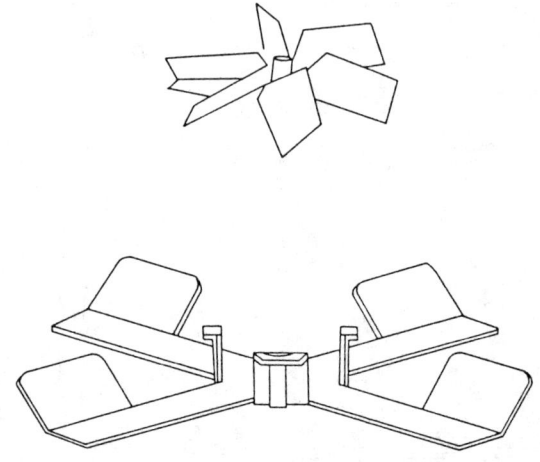

Fig. 3.6 Mode stirrers

into food items such as roasts and the desired final internal temperature preset on the control panel. When the roast has reached the preset temperature the oven will turn off automatically. For ovens with a turntable, the probe design permits it to rotate freely.

In the absence of a temperature probe it is possible to measure product temperature with a dial type bimetallic thermometer and non-metallic thermometers filled with a non-toxic, harmless indicating fluid. Neither of these should be used in thermal microwave ovens. The plastic lens on the bimetallic thermometer will melt, and at high temperatures a liquid filled thermometer could break.

**Humidity sensors.** A few models have been offered with a humidity sensor (Fig. 3.7) located in the air exhaust stream. The sensor, which contains a moisture sensitive substance such as lithium chloride, measures the humidity of the moisture liberated when the object is heated and converts it to a measuring quantity that controls the power to the object being cooked (Risman, 1974).

**Gas sensors.** These respond to sudden emission of food aromas, much like smoke detectors respond to smoke, and program the remainder of the cook cycle.

**Programmed recipe cards.** Ovens with magnetic readers respond to the insertion of cards with coded cooking directions. The program may vary in both time and power as well as changes in these quantities and delays to allow some conduction heating for a specific quantity of a recipe. Experimental ovens have been equipped with recognition devices such as used at supermarket check out counters. In the future it is conceivable that heating directions on the food package will be printed in code form such as the Universal Product Code so that instructions can be programmed into the oven control circuit by means of a similar device.

# THE MICROWAVE OVEN

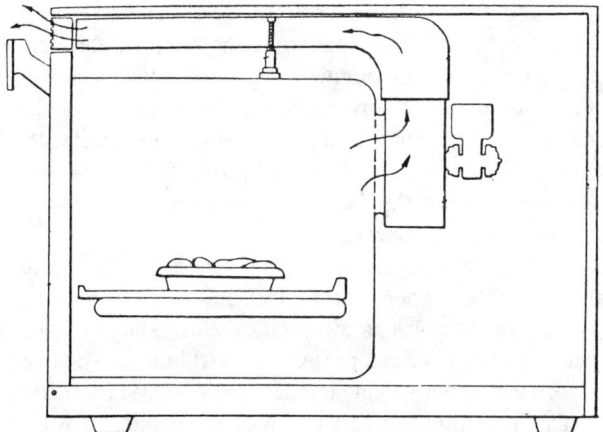

Fig. 3.7 Humidity sensor in microwave oven (Risman, 1974).

**Verbal directions.** Speech synthesis has been demonstrated in microwave oven controls so that a spoken command can be given to announce some recipe action that should be taken; for example, season, stir, turn roast.

**Variable power.** Power level control acknowledges the need to cook some foods more slowly than others in order to obtain the best cooking results. Thus high power is convenient for heating soups and beverages, while a somewhat lower power setting (60%) would give best results when heating refrigerated foods where some conduction heating helps to improve the cooking uniformity. It is difficult to prescribe a specific power level because there are so many food variables to contend with; e.g., shape, density, temperature, quantity. An even lower power level (30 to 40%) is best for meat roasts where the density and dimensions require slower cooking. The lowest setting (10%) is used for sauces and gravies to allow simmering to develop full flavor.

Defrosting frozen foods without the need for special attention can be carried out successfully at under 250 watts, although programmed heating, initially at higher power followed by completion at lower power will save considerable time. Although there may be some disparity among cookbooks these power levels are reasonable and will give satisfactory results.

Variable power is accomplished by varying the magnetron on-time as a percentage of a specific time base. Thus an oven with a 2 second time base provides 10% power by cycling the magnetron on for 0.2 seconds and off 1.8 seconds. The time base varies widely among ovens. The range is about 1.1 to 60 seconds. The starting time also varies among ovens. This is the time between turning the oven on and the delivery of microwave power. It can vary between 1 and 3 seconds.

**Sensing oven temperature.** Sensing the oven temperature also has been used to control microwave cooking. One microwave oven has a built-in analog computer that records the oven air temperature when the oven is energized. A blower system allows the air to follow a prescribed path over the food and through an opening in the top of the oven where its temperature is measured by a thermistor. As the food cooks, the temperature of the air flowing past the thermistor rises slowly. Since the difference in air temperature from the onset of cooking to the time the food reaches the appropriate serving temperature has been programmed into the computer, the oven power will cut off when the thermistor signals to the control circuit that the proper temperature has been reached.

The automatic sensor is normally used for foods of similar composition and of the same size and quantity. The control panel on one such oven has five settings for one to five servings or portions of specific items. The user merely selects the number of portions, up to five, and presses the start button. The oven will turn off automatically when the proper serving temperature has been reached. The automatic sensor in this case is used only when the oven is operated at full power.

Foods that can be cooked using this feature include beverages, meats, fish and vegetables, frozen vegetables in cartons or pouches, casseroles and cake mixes. It may also be used for certain cooking operations requiring user attention such as scrambled eggs. In the case of eggs, the sensor cuts off power at the time the eggs should be mixed, and completes the cooking on the second cycle.

**Weight sensing.** So called "Smart Ovens" represent another innovation in oven control. The user supplies information about type of food to be cooked (meats, casseroles, fish, cake, etc), its present condition (frozen, refrigerated, room temperature) and the degree of cooking desired. The oven automatically determines the best cooking procedure and begins the cooking cycle. The key to this oven's "smartness" is sophisticated microwave processor control and an ingenious weight sensing system. The oven has a built-in scale that transfers the weight to a single point. A dual strain gauge is used that weighs to the nearest 20th of a pound.

A new technique has recently come along that makes weight sensing alone obsolete. It is called "fuzzy logic."

**Fuzzy logic.** Fuzzy logic technology is making it possible to produce very clever microwave ovens. Sharp Corporation demonstrated a microwave oven with 11 separate sensors at the Consumer Electronics Show in Chicago in the Summer of 1991. The sensors monitor aroma, humidity changes, food weight, shape, thickness and quantity, oven temperature, and can detect if the oven is being operated empty. This oven, in addition to being a 700 watt microwave oven, is a forced convection oven with a capability of operating at up to 570 F, and has a ceramic broiler element. To operate, one presses one of four touch pads on the control panel: microwave range, grilling, reheating or defrosting. In the first two modes the user can select from among 43 foods in 14 groups. This action activates the sensors that determine the food size, weight, quantity, shape and thickness. A microcomputer then carries out a computation in 7 seconds that sets the operating time and power automatically. As cooking progresses the sensors up-date the program and make necessary adjustments to power and time. In addition, the oven has a turntable mounted on the oven floor that can be replaced with a plastic device to knead bread dough prior to baking, mash potatoes, cook stews and puddings, and make jams and jellies to name a few. Making mashed potatoes provides a good example of how fuzzy logic works. With the mixing bowl in place and whole or cut up potatoes in the bowl, a torque sensor on the mixing arm detects the resistance offered by the potatoes. In conjunction with other inputs such as humidity and optical sensing that indicates the presence or absence of pieces of potato, the speed of rotation of the arms either increased or decreased. Fuzzy logic takes these multiple inputs and makes decisions based on that input. In a sense fuzzy logic simulates the human reaction of cooking by inspection (Daniels, 1991).

## MICROWAVE OVEN TYPES

### Microwave ranges

Since the first combination microwave oven was introduced in the mid-1950s by the Tappan Stove Company, a microwave oven with a so-called 'browning element,' essentially a broiler, there have been a variety of combinations of microwave and thermal energy ovens introduced. The majority of the combination ovens are the microwave range models that, for the most part, are purchased to replace the conventional range when a kitchen is being refurbished and to a greater extent they are being installed in new homes and apartment houses. The microwave range is a conventional oven with microwave power and the usual surface heating elements, gas or electric.

### Microwave convection oven

The microwave forced convection oven has three modes of operation: microwave, forced convection and the simultaneous use of microwave and forced convection heating. Convection is provided by means of a high speed blower moving high temperature air across the food being cooked. The combination of microwave and forced convection heating gives typical surface browning and flavor development in microwave time.

### Microwave impingement oven

A variation on the forced convection principle is provided by a technique known as impingement heating. As shown in Fig. 3.8, high velocity air is directed at the product instead of being simply moved around the oven, thus providing much more effective heat transfer. Movement of the jets or the product is required for uniform browning, otherwise a pattern of brown spots like freckles will result. This technique has been used in conveyorized heating of plated meals in combination with

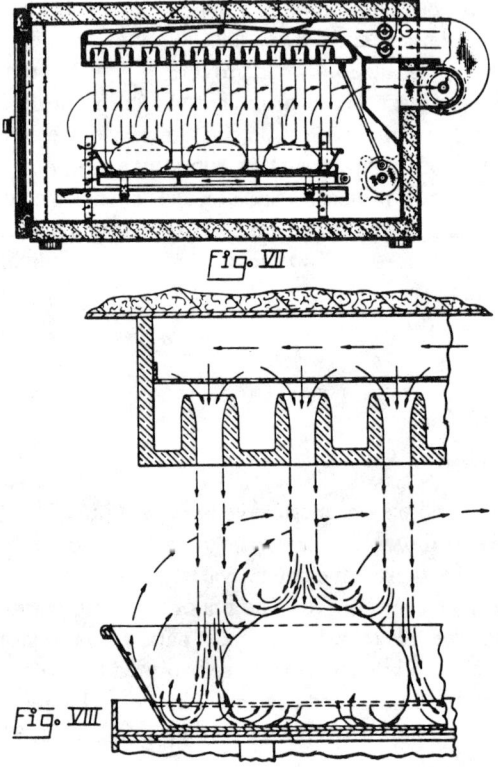

Fig. 3.8 Jet impingement oven (Smith, 1975).

microwaves in hospital foodservice and in a recently demonstrated vending machine for french fried potatoes, fried chicken and pizza. Impingement heating alone is used in hundreds of conveyorized pizza ovens in pizza parlors throughout the world.

## Higher powered ovens

In June 1983, Amana Refrigeration, Inc., introduced a consumer microwave oven with 1000 watts of microwave power. A reduction in cooking time of 23 to 38% over the usual 600 to 700 watt microwave ovens was claimed. This oven requires a dedicated 20 ampere circuit. The oven was taken off the market as being non-competitive.

Microwave oven development by the Philips Company in Holland in the 1960s was based on a 1.2 kW magnetron because it was available at the time. However, it was felt that the Dutch market would need the extra power to prepare their more hearty meals.

**Low powered ovens**

Although low powered microwave ovens at around 400 and 500 watts have been in common use in Japan for many years, they became available and in general supply on the United States market only around 1985. Korean oven manufacturers have become the major suppliers of low powered ovens. Because of the much lower cost, some at even less than $100, sales grew rapidly until they now represent perhaps 35-50% of all microwave oven sales. Heating at 400 and 500 watts understandably is slower than at 600 to 700 watts and this adds another variable for the product developer.

## STANDARDIZATION OF MICROWAVE OVENS

One of the major headaches facing the food product developer for the microwave oven market is the absence of microwave oven standards. No two oven models are exactly alike. For the most part they differ in total power available and in the power level associated with the variable power settings in most ovens. In the absence of standards, food product developers must carry out heating tests in a variety of microwave ovens and provide heating directions based on a range of heating times. Ovens should be selected with and without turntables; with several different cavity sizes and power levels; and other features that might affect heating results such as the method of microwave energy feed into the cavity. With such a precisely controllable heating device as the microwave oven this should not be so. However, this situation can be expected to continue for some time.

There is, as one might expect, an on-going effort through the Cooking Appliance Section of the International Microwave Power Institute to develop and agree upon power setting standards and technology. Many ovens have as many as ten power settings. Some microwave oven interests feel that this is not necessary and that five settings can cover most of the needs of oven users. If some agreement can be reached this could lead to the ultimate in microwave oven control in which coded messages written on food packages containing precise cooking or heating instructions could be read by recognition devices, such as used at supermarket check-out counters, and transmitted to the microprocessor oven control. This would eliminate all judgment settings and insure reproducible results.

## MICROWAVE OVEN SAFETY

Safety is perhaps the most misunderstood fact about microwave ovens, largely because it has been misunderstood by those capable of giving widespread publicity to inaccuracies about the subject. Many newspaper columns, some by syndicated columnists, TV programs, and at least one book and two articles in The New Yorker magazine by the same author came down hard on microwave ovens based on misinterpretation of the facts. In the face of what some might consider irresponsible journalism, truth won out; truth backed by an impressive safety record.

In the minds of many, radiation is identified with X-rays, atoms bombs and nuclear power. But radiation is much more than that. There are forms of radiation that bear no relationship to nuclear radiation. They include radio, television, microwaves, infrared radiation, and everything in between. We obtain great benefits from these energy sources and for the most part are safe from excessive exposure to them. We could not exist without the sun's radiation, but even the sun's energy can be harmful if taken in large doses: witness the concern about skin cancer from excessive exposure to the sun's rays. The kitchen electric range elements are a familiar source of infrared radiation to all of us and unless we touch one of the hot elements there is little likelihood that we will come to any harm from using such a range.

Microwave energy used in microwave cooking is confined inside the microwave oven where its energy can be put to good use. Performance standards and design criteria have been extablished by law so that leakage or emission of microwave energy from these ovens remains at extremely low, safe levels. Considering that there are many millions of microwave ovens in use today, and many have been in use for many years, there is not one documented case of anyone having been harmed from microwave energy leakage from these ovens.

Of greater concern is the matter of safety from food temperature hazards and from steam. The Consumer Products Safety Commission has recorded numerous instances of steam burns suffered by children opening a bag of freshly microwave popped corn, as well as from spills of hot food being removed from an oven. Pressure is being placed on food manufacturers to provide safe handling instructions on food packages and to encourage adult supervision of young children using microwave ovens.

**Emission standards**

Public Law 90-602, the "Radiation Control for Health and Safety Act of 1968," was passed on October 18, 1968 (Anon, 1970a). One of the purposes of this law was to establish performance standards for various electronic products' radiation so that the public could not be exposed unnecessarily to excessive amounts of various kinds of radiation given off by these products. Items such as television sets, sun lamps, microwave ovens and a list of many other products are covered by this law.

The performance standards for microwave ovens eventually went into effect on October 6, 1971. Under these standards, microwave emission from microwave ovens cannot exceed one (1) milliwatt per square centimeter ($mW/cm^2$) measured two (2) inches from the oven prior to its sale: and five (5) $mW/cm^2$ over the life of the oven. The measuring equipment to be used is also specified in the standards.

To this should be added some commentary on leakage monitors of which a number of very inexpensive devices have been offered for sale. Many of these have been evaluated by the Bureau of Radiaological Health of the Food and Drug Administration and shown to be less than satisfactory (Anon. 1970b). BRH provides oven checks with the proper instrumentation if requested. Companies that service microwave ovens are also capable of providing this service, though there may be a charge. It should be pointed out that the agency responsible for field testing microwave ovens discontinued the practice because they found that ovens rarely leaked above the standard unless they were abused or serviced improperly (Barron, 1991).

## Exposure standards

Long before microwave oven emission standards were established, industry had operated with a voluntary standard of 10 mW/cm$^2$. This was an exposure standard that had been established in the 1950s to protect technicians working around radar equipment from being exposed to possibly dangerous levels of microwave radiation. Animal experiments had demonstrated that exposure at levels of 100 mW/cm$^2$ or higher could be harmful over a prolonged period of time and concluded that a safety factor of 10 would be provided by setting an exposure standard at 10 mW/cm$^2$.

The difference between "exposure" and "performance" standards was where the press had gone astray; that is, failing to recognize the difference between the two. The "exposure" standard is based on "whole" body radiation; that is, a condition in which the entire body surface is receiving radiation. Another way to express this is to say an individual is immersed in a microwave bath in which each square centimeter of the body surface is exposed to a heat load of 10 milliwatts. Taking a hypothetical body surface of one square meter (10,000 cm$^2$), the total heat load amounts to 100 watt-hours, which the average body's cooling system can handle. By way of contrast solar energy reaching the Earth on a sunny day may be 60 to 100 mW/cm$^2$. One can receive a nasty sunburn from such radiation and many do.

The "performance" standard insures much less, almost infinitesimally less, exposure. The reason is perhaps more clearly shown in the illustration below (Fig. 3.9). Radiation diminishes by the square of the distance from the source. A meaningful example can be provided by standing in front of a fire in a fireplace. Up too close, the heat is unbearable. At the right distance, the sensation of heat is pleasant. All radiation behaves in the same way. A microwave oven emitting 5 mW/cm$^2$ measured two inches from the oven will provide only about 0.05 mW/cm$^2$ at an arm's length. Using the same relationship, if the meaurement at the oven door were 100 mW/cm$^2$, the level at an arm's length would be about 1 mW/cm$^2$. It is also important to realize that this is "partial" body radiation; that is, only the part of the body facing the oven, not "whole" body radiation and so the total heat load is much less. The pretigious Journal of The American Medical Association in a 1971 editorial shortly after the oven standards were promulgated pointed out that the standard carries with it a safety factor of 10,000. Many other authorities have also stated strong personal convictions that the U.S. microwave oven standard is quite conservative and under it microwave ovens are safe. A statement by Dr. James Van Allen,

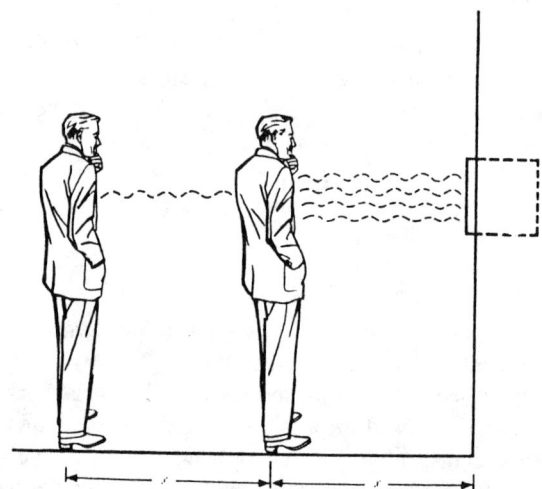

Fig. 3.9 Illustration of microwave energy leakage level diminishing by the square of the distance from a microwave oven.

the noted radiation authority, has been quoted frequently: the microwave hazard from operating a microwave oven is "about the same as the likelihood of getting skin tan from moonlight." The list of authorities includes some from Eastern bloc countries whose stricter exposure standards have often been confused with U.S. oven performance standards in statements such as "the U.S. standard is 500 times higher (allows 500 times more radiation) than the Soviet standard."

The relationship between the exposure standards and typical oven operator exposure is shown in Fig. 3.10. It can be seen that the levels used in microwave diathermy are well above any of the exposure standards. The exposure limits for oven operators shown in the figure are based on an oven emission level of 5 mW/cm$^2$, although this level is considerably above the average. Typical operator exposure at 3 feet for 1 minute from a microwave oven at the above emission level is about 10 times below the Eastern European standards.

By any criteria, operation of microwave ovens pose no threat to the health and safety of the operator from microwave radiation.

## MICROWAVE OVENS OF THE FUTURE

Future oven designs may be dictated more by food preferences than any other considerations. There are strong indications that greater usage of prepared, portioned foods in disposable packages will be the norm and that prime cooking of such large

THE MICROWAVE OVEN 83

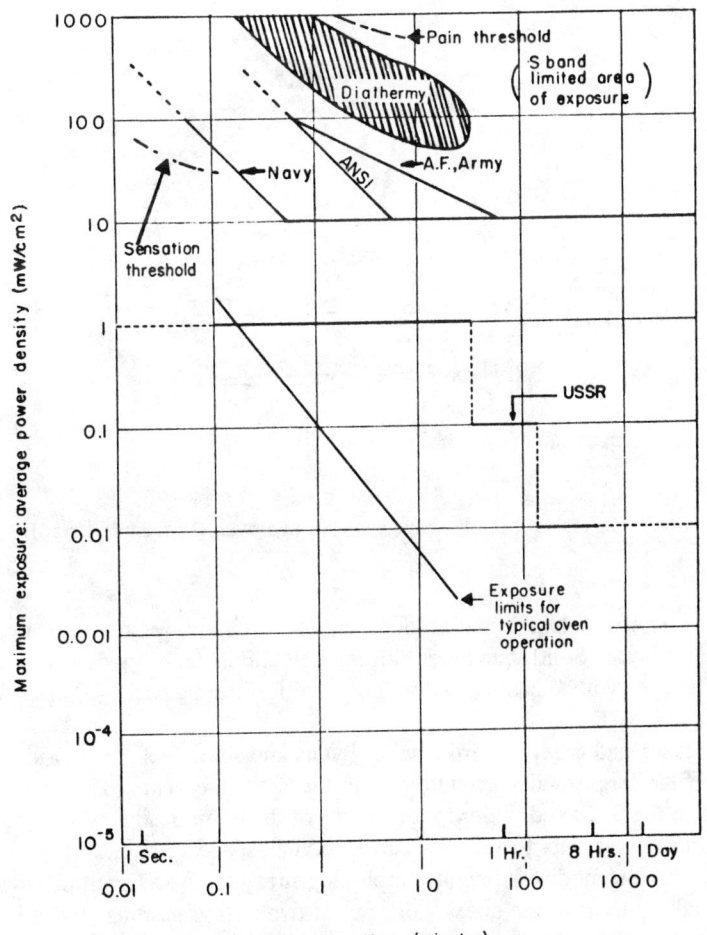

Fig. 3.10 Comparison of exposure levels of microwave energy (From Osepchuk, 1978 with permission).

food items as meat roasts will be the exception. This would suggest a change from using large microwave ovens to much smaller ones. This change coupled with sophisticated solid state microwave generators and power supplies means that microwave ovens will not need to be much larger than the packaged food products placed in them to be heated. An early example of a unique microwave oven design that was years ahead of its time is shown in Fig. 3.11. A microwave oven in a drawer also is a distinct possibility (Decareau, 1979).

A precursor of the oven in a drawer might have a viewing window in the counter surface so that cooking or heating can be observed easily. Opening the drawer, which will be at the proper height, determined ergonomically, will enable the user to stir,

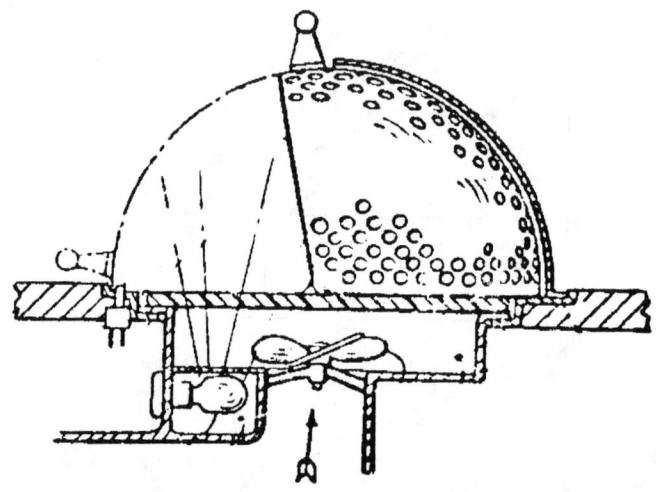

Fig. 3.11 Novel dome shaped microwave oven (Woodman, 1958).

season or otherwise handle the product without having to remove it from the oven as is now the case. Solid state mode shifting will eliminate the mode stirrer and turntable. The oven will be completely programmed for heating packaged prepared foods as suggested earlier.

There is a trend today toward smaller ovens and some feel that small ovens will represent the largest sales growth area in the next five years. The price of these ovens is in the $150 and under range. Some of them are designed to be suspended below kitchen cabinets to release scarce counter space.

Some very recent developments employing fuzzy logic and multiple sensors are already taking most of the guess work out of oven programming. Indeed, the programming is automatic to a large extent and self adjusting to upgrade the time and power settings as cooking progresses. Marketing of one such oven began in July, 1991 in Japan at a hefty $1000 each. It is expected that cost reductions will occur and the oven will be imported to the United States in a year or so. There is always a market for some top of the line appliances and microwave ovens are no exception.

Microwave ovens also are being used on recreational vehicles and pleasure boats. A recently introduced luxury car has been equipped with a microwave oven with power provided by an alternator off the engine.

## SUMMARY

The microwave oven offers unusual convenience, speed, flexibility and nutritional advantages over other methods of heating foods. The developer of food products for this market would do well to reflect on all these advantages and oven features and strive to develop products that match or mate with these advantages and features. This is particularly important since most surveys indicate that the main use of the oven is to heat prepared foods: frozen, chilled, and shelf stable.

A primary consideration should be given to designs that will minimize hazards to young children who may operate the oven. Thus, the food product developer must keep safety in mind in developing the food package.

## REFERENCES

Anonymous (1970a). Performance standards for microwave ovens. Fed. Reg. *35* (194) 15642-15643.

Anonymous (1970b). A comparison of microwave detection instruments. BRH/DEP 70-7. U.S. Government Clearinghouse for Federal Scientific and Technical Information, Springfield, VA 22151.

Barron, J. (1991). Personal communication. Dept. of Health & Human Services, Rockville, MD.

Berntsen, W.T. and David, B.D. (1975). Determining the electric field distribution in a 1250 watt microwave oven and its effect on portioned food during heating. Microwave Energy Appl. Newsl. *8* (4) 3-10.

Bobeng, B. and David, B.D. (1975). Identifying the electric field distribution in a microwave oven: A practical method for food service operators. Microwave Energy Appl. Newsl. *8*(6) 3-6.

Copson, D.A. (1975). Microwave Heating, 2nd Edition. Van Nostrand Reinhold/AVI, New York.

Daniels, J. (1991) Personal Communication. Sharp Corporation Mahwah, New Jersey.

Decareau, R.V. (1979). Microwave oven features and accessories. Microwave Energy Appl. Newsl. *12* (6) 16-18.

Decareau, R.V. and Peterson, R.A. (1986). Microwave Processing and Engineering, Ellis Horwood Ltd., Chichester, England, 224 pp.

IEC. (1988). Methods for measuring the performance of microwave ovens for household and similar purposes. International Standard, IEC 705. International Electrotechnical Commission, Geneva, Switzerland.

Kumpfer, B.D. (1972). A simple system of microwave pattern measurements. Microwave Energy Appl. Newsl. *5* (4) 10.

Osepchuk, J.M. (1978). A review of microwave oven safety. J. Microwave Power *13* (1) 13–26.

Peterson, A. and Foerstner, R.A. (1971) Evaluation of microwave oven cooking performance. Microwave Energy Appl. Newsl. *4* (1) 3–8.

Ringle, E.C. and David, B.D. (1975). Measuring electric field distribution in a microwave oven. Food Technol. *29* (12) 46, 48, 50, 52–54.

Risman, P.O. (1974). Method and device for producing heating of moisture-containing objects. U.S. Patent 3,839,616.

Smith, D.P. (1975). Cooking apparatus. U.S. Patent 3,884,213.

Staats, J.E. (1982). Microwave oven with novel energy distribution arrangement. U.S. Patent 4,354,083.

Stanford, M. (1990). Microwave oven characterization and implications for food safety in product development. Presented at the 25th Microwave Power Symposium, Aug. 27-29, Denver, Colorado.

Thorsell, U. (1991) Personal communication. The Swedish Food Institute, Göteborg, Sweden.

Wilhelm, M.S. and Satterlee, L.D. (1971). A 3-dimensional method for mapping microwave ovens. Microwave Energy Appl. Newsl. *4* (5), 3–5.

Woodman, K.L. (1958). Electronics Ovens. U.S. Patent 2,956,144.

# CHAPTER FOUR

# PACKAGING FOOD PRODUCTS FOR THE MICROWAVE OVEN

## THE FUNCTIONS OF PACKAGING

The package for heating foods in microwave ovens serves several important functions and is somewhat more complex in design than a package for heating in a conventional oven. The microwave package

- protects the product in storage and distribution,
- sells the product,
- controls the heating of the product, and
- may function as a serving dish.

This last function may not be essential to performance, but there is a growing tendency toward packaging foods in disposable serving containers to avoid the need for dish washing. There is also a concerted effort to produce attractive, functional containers at affordable prices. Such containers also help to sell the product. Now we are asked to consider the negative effect of disposable packaging on the environment and to do something about it. Some discussion of each of the above functions is in order.

**Product protection**

Although there is a tendency to think only in terms of frozen foods when considering new product development for the microwavable foods market, the needs of the market actually are much broader. Prepared frozen foods is a huge market, a still growing one, and offers many opportunities for new product development. But there is also a market for chilled foods, shelf stable foods, dry mixes, dehydrated foods, baked goods, desserts and specialty foods. Thus it is necessary to consider packaging for all kinds of food products as well as containers for the contents of the packages. When we speak of packaging here we are in fact speaking of packaging and serving containers and in some instances, they can be one and the same.

Almost any material that has Food and Drug Administration approval can qualify for microwave heating or cooking applications. Thus glass, glass-ceramics, ceramics, plastics, paper, and metals are all candidate materials.

**Glass.** Glass is made from non-metallic earths, basically silicates (sand). It is nonporous and does not absorb moisture or food materials. Although it is generally known that glass containers can be used in microwave ovens with impunity, there has been, until recently, no substantive body of data in the literature on this subject. Gerling (1981) reported that ordinary soda lime glass was reasonably transparent to microwave energy and also able to tolerate a substantial pressure build-up if inadvertently heated without removing the closure. Several types and sizes of closures were tested, including reclosures to simulate re-use by consumers where food left in the seal area acted as a kind of cement. Normally, the closure deformed, releasing the pressure. Pressures up to 56 pounds per square inch were recorded. Breakage of the glass appeared to be an unlikely consequence of pressure build-up in microwave heating of food in glass jars. Even glass jars with faults typically encountered during manufacture were strong enough to retain their integrity until the closure deformed.

Anchor Hocking has developed a closure for glass jars that vents automatically. As pressure builds up, the seal breaks. The seal does not have to be removed from the jar before the jar is placed in the microwave oven. Called Micro-Serve™, jars sealed in this manner are retortable. The companion container was developed by Owens-Brockway Company. The closure consists of a metal disk bonded with polypropylene to protect the metal edge. Plastisol lines the inner circumference to give a good seal. Consumer concern about metal in the microwave oven was taken into consideration in the design. Damaged closures with metal exposed did not arc in the oven. The 77 mm press-on top closure can be used with a variety of different containers. The closure snaps back on to protect leftovers in storage.

Heat resistant glass containers are recommended for microwave oven use. Usually such containers or dishes have been tempered to relieve stresses and strains, and carry labels such as "Heat resistant" or "Ovenproof" or trade marks such as "Fireking" and "Pyrex." This kind of glass can be used in conventional ovens as well as combination conventional and microwave ovens. In addition such glass can be used for freezing foods and can be taken from freezer to oven without danger of breakage.

**Glass-Ceramics.** Glass-ceramic dishes are made like glass, but from special ingredients, then heat treated. Like glass, they are non-porous and do not absorb moisture or food. There are two types of glass-ceramics: glazed and unglazed. Glazed dishes are not recommended for microwave oven use. The dish and glaze have different coefficients of expansion. The dish contains materials that cause it to heat faster in a microwave oven than the glaze. This difference in heating rates may cause such dishes to shatter. Unglazed glass-ceramics, on the other hand, are even more heat resistant than heat resistant glass and are highly recommended for microwave oven use. Sold under the trademark, "Corningware," this glass-ceramic product can be used not only in microwave ovens, but also conventional ovens, on range tops and under broilers (Fig. 4.1).

Fig. 4.1 Example of glass-ceramic microwave cookware (Courtesy of Corning Glass Co, Corning, NY).

**Ceramics.** This term covers a variety of products including pottery, earthenware, fine china and porcelain, and these products vary in microwave absorption characteristics. They are not fired at as high a temperature as glass and glass-ceramics, hence the moisture content is higher and this accounts, in part at least, for the greater microwave absorbency. In general, they all can be used in microwave ovens, though some will be less suitable for long heating cycles because they compete with the food on them for the available energy. Some ceramic cups and dishes may become too hot to handle before the contents have reached serving temperature.

Although glass, glass-ceramics and ceramics would not appear to be candidate materials for convenience food packaging, it has been said that tempered glass might be producible in a TV-dinner configuration at near competitive pricing. Some years ago a mass producible ceramic dish that could be recycled into new dishes by grinding and recasting was proposed, so this is not a new concept. With the current concern about environmental waste, regulations may one day require that all packaging be recyclable, reusable or returnable for reuse.

**Plastics.** Snedeker and McKenna (1978) evaluated a number of plastic materials for suitability as microwave cookware materials. Criteria included safety factors, dishwasher performance, impact resistance and food odor retention. The list of materials included polysulfone, filled thermoset polyester, modified poly (phenylene oxide), poly (butylene terephthalate), polycarbonate, polypropylene and poly (methylpentene).

Microwave suitability was judged by the somewhat controversial "full power-no load" test in which the material was exposed to full power (600 watts) for two hours or until failure occurred. This test was controversial because such severe exposure conditions, if accepted as a standard, would eliminate many perfectly suitable materials. It is interesting to observe that in some microwave ovens the Pyroceram™ floor can reach temperatures of 450 to 500 F during such no load testing.

Other data obtained included heat deformation, odor pick-up and retention, auto ignition, FDA status, impact resistance and dishwasher exposure. Sources of odor pick-up were Worcestershire sauce, cloves, onion sauce and pickle juice. Of materials tested for odor retention only polysulfone retained no detectable odor after two days standing. Polysulfone was found to be superior to the other candidates in some functional respects with no evidence of any shortcoming in any performance category.

A parallel series of tests made by a consulting firm for a major material supplier compared polysulfone, a thermoset polyester, a thermoplastic polyester, and poly-4-methyl pentene-1. The results of this comprehensive study were presented at a microwave oven workshop in Atlanta (Stehle, 1979). A variety of food items were used to test the reaction to high fat temperatures, high sugar temperatures, protein stain or residual, tomato staining and small oven loads. Bacon was used for high fat temperatures; peanut brittle for high sugar temperatures; beef patties and meat loaf for protein residual; and pork roast for high fat and protein residual. Appropriate utensils and accessories made from these plastics were used. In addition to performance, they were evaluated also for cleanability.

In brief, the study concluded that (1) cookware made from polysulfone and thermoset polyester can be used with any food and for all microwave cooking purposes; thermoset polyester was more difficult to clean in some cases, but was usable; and (2) cookware made from thermoplastic polyester and methyl pentene isomer showed limitations in some of the shapes tested, but were satisfactory for many foods. These two were the only containers that warped or became distorted. The thermoplastic container usually became permanently distorted, while the methylpentene isomer recovered its shape on cooling.

Colato (1978), in reviewing the characteristics of a number of packaging materials for their suitability for microwave oven usage, referred to a less severe test than the no-load test. This test involved heating the material for one minute in a microwave oven at 600 watts with one-half cup of water alongside. Suitable materials should be cool to the touch after this exposure. Hot materials should be suspect and if too hot to handle should be ruled out for microwave oven usage.

Colato (1978) listed other plastic materials that may be used in microwave ovens. The list included: polyethylene, polypropylene, acylonitrile, butadiene styrene (ABS), polycarbonate, nylon, styrene acrylonitrile (SAN) and mixtures of styrene and acrylic sold under the trade name Dylark.

Polyethylene, though rigid, distorts at temperatures around 170 F and is brittle at low temperatures. Polypropylene has a higher distortion temperature, around 230 F, is not rigid but has a tendency to become brittle. ABS can tolerate temperatures in the 190 to 220 F range depending on the grade. It is somewhat subject to damage by abrasion. Polycarbonate has a distortion temperature near 250 F and is extremely break resistant. Nylon is strong, but has a low tolerance to temperatures at 150 to 170 F. SAN is strong, but limited to 190 F. Styrene and acrylic mixtures, though tolerant to temperatures up to 230 F, are somewhat brittle.

Expanded polystyrene, although it has good food heat retention properties, has not been completely successful in microwave oven usage. Although transparent to microwave energy it is not immune to high temperatures attained by some food products. Monte and Landau-West (1983) tested a wide variety of frozen foods heated in expanded polystyrene containers. The foods were heated to 180 F and held at this temperature for 10 minutes. Any container leakage, major distortion or softening that might result in container failure resulted in a "not recommended" rating for that product. Fatty foods (such as fried foods, gravies, certain cheese sauces, fatty meats and buttered foods) gave high failure rates. Foods with much gravy showed breakdown of the container at the gravy line.

In triangle tests in which the foods used were baked beans, cream sauce, mashed potatoes and stuffing mix heated in glass and expanded polystyrene, only in stuffing mix did a significant number of panelists detect a difference. Although the failure rate was highest for certain fatty foods there appeared to be a difference between saturated and unsaturated fats. Chili made with corn oil (10% and 15%) caused leakage when cooked to 180 F, whereas chili made with beef tallow had no effect on the container.

Thermoset filled polyester dishes were used for the first time more than 20 years ago for certain precooked frozen hospital diets. The plated foods were covered with a sheet of aluminum foil and capped with a thin polyethylene dome. According to directions supplied with these diets, for conventional oven heating the dome was removed. For microwave heating, the foil was removed and replaced with the polyethylene dome. Later, dinners packaged in this way were offered in supermarkets. Even more recently several lines of frozen dinners were introduced to the retail trade. Since these dinners were priced well above TV-dinners, the cost of the filled polyester dishes, around 15 cents each, would seem to be easily tolerated. Sales, however, have fallen off partly because of consumer resistance to the high price of such dinners and because a thinner, less expensive plastic compartmented plate has replaced the filled polyester dish.

## Barrier plastics

Although we have spoken mainly about packaging prepared foods for the microwave oven user, it is conceivable that some of these foods will be microwave processed in the package that the consumer ultimately will place in a microwave oven to heat for serving. In-package microwave pasteurization and sterilization are two processes that are receiving considerable attention today. Both offer greater convenience than frozen foods. They also are cheaper to market since the costs of freezing, freezer storage and distribution, and display in freezer cabinets are eliminated. There are indications that the differences in quality can be quite small. Microwave in-package sterilization has become possible owing to developments in barrier materials that are able to tolerate the higher processing temperatures and that can assure a suitable shelf-life.

Hecht and Haskell (1976) reported on barrier films for shelf-stable packaging that consisted of a metallic phosphate sandwiched between a polyester film and a heat sealable layer. The film was said to be 100 times better in terms of oxygen transmission rate than the best transparent films then available. The barrier coating, however, did not stand up to retort processing conditions.

Formerly, the barrier function was accomplished with a layer of aluminum foil sandwiched between two layers of polyethylene. The foil layer, however, prevented the package from being microwavable. Currently, the barrier function is being handled by a layer of ethyl vinyl alcohol (EVOH) or polyvinylidene chloride (PVDC) sandwiched between layers of polyethylene (PE), polypropylene (PP) or high impact polystyrene (PS), although this packaging also loses some of its barrier properties when exposed to water and high temperatures. Vinylidene chloride/methylacrylate polymers are said to provide twice the barrier properties of EVOH and PVDC (Lyons, 1988).

Glass-coated polyester containers recently reported are clearly one of the most innovative developments in packaging to appear in years. QLF (for quartz-like film) jars have a thin coating (less than 2000 angstroms) of silicon dioxide. This PET/glass combination provides a three to 10 fold improvement in oxygen barrier over uncoated PET containers, and they are microwavable and compatible with recycling processes.

Lamipac, Metal Box, Ltd.'s retortable containers are thermoformed from coextruded PP/PVOH/PP with a foil/PP flexible heat sealed lid. Cobalplast, Continental Can Co.'s retortable system, consists of 2-8 layers of coextrusions of PE, PP and high impact PS with PVDC or EVOH.

Thermoformed containers of crystalline polyester terephthalate (CPET) are an alternative to the multilayered constructions. Most prepared food packers purchase preformed containers, however, a few large firms are capable of manufacturing their own containers directly from coextruded sheet stock. Lidding for foods to be retorted for shelf-stability is usually heat sealable aluminum foil.

Equipment such as that shown in Fig. 4.2 is available from a number of manufacturers to form containers from roll stock and, after filling, to seal them. Such equipment, for example, is being used in packaging fresh pasta and a wide variety of meals that are to be microwave pasteurized.

Some prepared products for the microwave oven are on the market using modified atmosphere packaging to extend their refrigerated shelf-life. Lidding is typically polyester film such as DuPont's OL Mylar heat sealed to the tray.

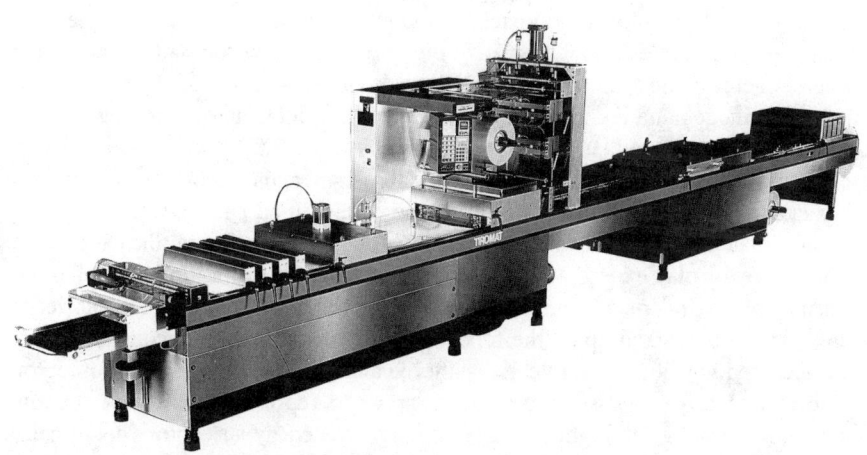

Fig. 4.2 Form, fill and seal packaging equipment (Courtesy of T.W. Kutter, Inc., Avon, MA).

## Frozen food packaging

CPET trays are functional in the temperature range from $-40$ to 220 C thus they are dual-ovenable. They are about twice as costly as aluminum foil trays. In spite of cost they have caught on because consumers perceived products packaged in CPET as being higher quality. The Campbell Soup Company's line of Swanson frozen dinners was one that was being converted over from foil to CPET. For frozen dinners foil is used as a hood and the consumer is instructed to remove the foil and return the tray to the carton for microwave heating. Polycarbonate is also said to be gaining favor for dual-oven applications. Trays made from General Electric's Lexan™ polycarbonate resin coextruded with polyetherimide resin were in commercial use for a frozen food product (Rice, 1986a).

Dual ovenable refrigerated entrees, which at least for the time being are shipped frozen to supermarkets where they are thawed and sold from the refrigerated display case, also are using CPET trays.

## Effect of plastic packaging on flavor

Some concern has been expressed concerning the effect of some plastic packaging materials on food flavor. Laperle (1988) pointed out that all packaging materials will alter the flavor of a food to some degree. He added that the flavor transfer potential of a food grade plastic can vary greatly and that the flavor problems can occur even though the plastic meets the extractable limits established by the Food and Drug Administration. Risch (1988) noted that the Code of Federal Regulations 1987 "does not specifically cover microwave packaging materials but does generally cover materials that are to be used up to or slightly above the temperature of boiling water."

It might be necessary for a food package manufacturer to prove that volatile compounds do not migrate from the package into the food. If migration does occur then the manufacturer must be able to show that the level is less than 10 ppb for any known carcinogen and less than 50 ppb for any other not Generally Recognized As Safe (GRAS) compound. In the case of microwave sterilized food products where the processing temperatures are 250 F or possibly higher, migration of volatiles from the package into the food might occur. In the case of food packages incorporating susceptor materials (see Chapter 5) to enhance browning and crisping, the temperature of the susceptor may exceed 400 F. Usually some heat discoloration of the susceptor is observed. Under such high heat conditions some of the metallic susceptor compounds might be picked up by the foods. Although aluminum foil containers have been accepted for food use for years, sputtered aluminum such as used in susceptors to brown and crisp food surfaces would appear to represent a different situation altogether. Possibly the higher temperatures generated by susceptors could cause plasticizers in the laminated film coating to migrate into food.

**Paper.** Paper dishes may be made from formed or pressed coated paperboard or molded from pulp and coated with polyester. Polyethylene coated sulfate paperboard had been in use for many years in Europe for prepared frozen food packaging for both microwave oven and convection oven foodservice applications. Polyester coated sulfate paperboard containers were introduced into the United States in the mid-1970s and some half dozen or more firms were supplying containers to food processors of prepared frozen foods (Fig. 4.3). These containers are being promoted as dual ovenable; that is, they are usable in microwave or conventional ovens and have acquired a substantial following.

Molded pulp containers are somewhat more rugged by virtue of their greater bulk and unlikely to collapse with a full load of food. Both pulp and paperboard have reasonably good knife and fork resistance and are acceptably attractive. Paperware

Fig. 4.3 Paperboard microwavable packaging
(Courtesy of Westvaco).

will usually cost less than reusable ware, especially when the cost of dish washing is considered.

Coated paperboard is made by extruding a highly heat resistant polyester resin onto solid bleached sulfate board. To compete with aluminum foil from a cost effective point of view, paperboard must be able to function exactly as foil containers in order to eliminate the need for expensive capital investment for equipment. Techniques were developed in 1974 and 1975 to form the extruded material into functional nested trays.

Some major food firms have opted for dual ovenable molded fiber trays. These are custom produced by Keyes Fibre Company (Fig. 4.4). The very nature of the process for making these trays, that is, from a slurry of pulp fibers, renders it impractical for food firms to produce their own trays as they are able to do with foil or plastic stock. The cost is said to be about half that of dual-ovenable plastic trays. Their Ovenware II line made from recycled and recyclable fibers can be formed in a variety of shapes to meet the needs of food manufacturers. A foil hood and plastic dome are usually provided with instructions to remove the foil and replace the dome when microwave heating.

Fig. 4.4 Molded pulp microwavable containers (Courtesy of Keyes Fibre Company, Stamford, CT).

**Metal.** Materials to be truly universal in application must be usable in microwave as well as combination microwave and conventional ovens, and in conventional ovens. Even though the market heavily favors the microwave-only oven over the combination oven, a majority of U.S. households still have conventional ovens. Frozen food cases in supermarkets have little space for two versions of the same item. There is one material that satisfies all requirements quite adequately, and it has been available as long as microwave ovens have been available. The material is aluminum. At one time a very large percentage of the frozen prepared foods on the market were packaged in aluminum foil containers. In 1975 more than 150 million pounds of aluminum was used in the manufacture of 8 to 10 billion foil containers. The usage today in the early 1990s has been reduced to a trickle.

Aluminum has exceptionally good properties as a material for packaging and heating frozen food. Brandt (1979) listed these properties in familiar, non-technical terms. Aluminum is:

| | |
|---|---|
| Friendly to foods | Doesn't rust |
| Good moisture barrier | Doesn't melt or burn |
| Good light barrier | High insulation value |
| Good gas barrier | Perceived quality |
| Odor barrier | Reusable |
| Easily formable | Recyclable |

The controversy over the use of aluminum foil containers in microwave ovens grew in intensity following publication of a study sponsored by the Aluminum Association, Inc., and the Aluminum Foil Container Manufacturers Association in 1977 (Decareau, 1977) and became an issue at a number of seminars and workshops as well as in the trade press. Championed by the aluminum industry, acknowledged by scientists and technologists, it was opposed by producers of other materials, notably coated paperboard, based on the mistaken assumption that only microwave transparent materials could be used in microwave ovens. The recognition of the usefulness of foil containers seemed to have fallen on deaf ears as suppliers of convenience food products for use in microwave ovens, to a growing extent, turned to paperboard and plastic containers to reach this market.

Consumers invariably find in their microwave cookbooks the caveat, "check your warranty before using metal in your microwave oven," yet are instructed to use aluminum foil to shield various parts of foods to slow down the cooking. To compound the confusion one major microwave oven manufacturer introduced an entire line of metal appliances and utensils for use in microwave ovens. What is the consumer to think?

Most articles that discuss packaging materials for microwave oven use usually dismiss metal with the argument that since metal reflects microwave energy it cannot be used in microwave ovens. The truth is that because it reflects microwave energy it can be used in microwave ovens. Obviously that statement needs to be clarified. Any food item completely enclosed in metal cannot be heated with microwave energy. If, however, one side is exposed as, for example, in removing the foil cover from a TV-dinner, then the dinner will receive energy through that open side and will be heated. Because metal reflects microwave energy from the sides and bottom of such a container, the food around the edges will not be overheated as it would be in a microwave transparent container. It is not uncommon for frozen food heated in a microwave transparent container to be boiling at the edges and still frozen in the center. There are limitations to the use of metal containers in microwave ovens, but most of the retail sizes in which frozen foods are found could be used.

Another argument used against aluminum or any metal is that arcing will occur and the magnetron power tube will be destroyed. Both of these arguments have become less defensible in recent years, first of all, because a number of metal microwave appliances for use in microwave ovens were being marketed by one of the major microwave oven manufacturers. In the second place, it also has been demonstrated that the magnetron power tube can tolerate much worse conditions, such as no-load operation (operating the oven empty), without serious reduction in the tube's operating life. Arcing may occur when there is metal to metal contact, as in placing an aluminum container against the oven wall or against another container. Arcing is most likely to occur when heating frozen foods because such foods initially represent a poor oven load exposed to high electric fields. This would mar the oven wall where the contact occurred, but would not otherwise damage the oven. The same could happen with metal appliances in a microwave oven.

A good practice when heating frozen foods in aluminum containers is to remove the foil lid and return the container to its carton as shown in Fig. 4.5. The carton not only prevents unintentional contact of the container with the oven wall, it also serves as a lid to contain steam which contributes heat to the food and prevents food spatters from soiling the oven.

A development of the Aluminum Company of America (Rice, 1984) was introduced in an effort to counter the metal in microwave oven phobia. The 7/8-inch deep, round, three-compartmented aluminum tray is coated with a double-layer vinyl-matrix and epoxy coating and lidded with a fitted plastic dome. The dome covers the rim of the tray thus preventing contact of the tray with the oven wall so that arcing cannot occur. The dome also provides freezer protection for the contents of the tray, aids moisture retention and promotes temperature uniformity during microwave heating. The tray also can be used in conventional ovens.

Frozen, prebaked fruit pies (Mrs. Smiths™) also were being marketed in round, coated aluminum pans (Wave'r Bake™ by Alcoa). Directions were provided for microwave defrosting the pies in these pans. Although a few cents more expensive than plain aluminum they were 30 to 40% less expensive than comparable CPET containers.

The latest in microwavable packaging is steel cans. Weirton Steel Corp. has developed a shallow can with a pull top lid that can be used for single servings of soup and probably many other foods. The cans are already in test market in Europe

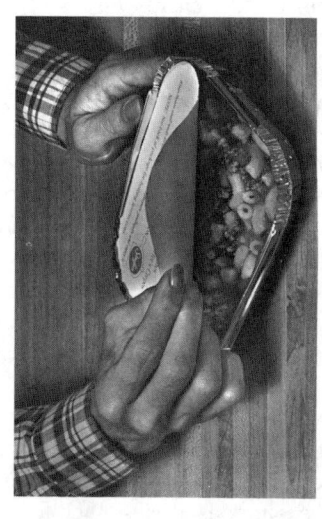
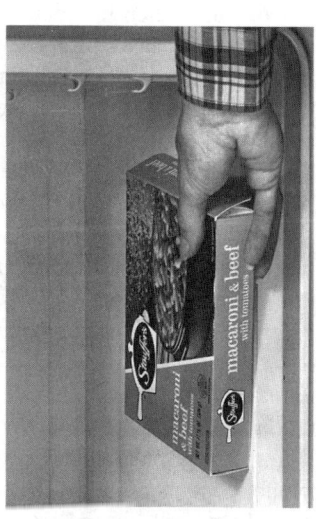
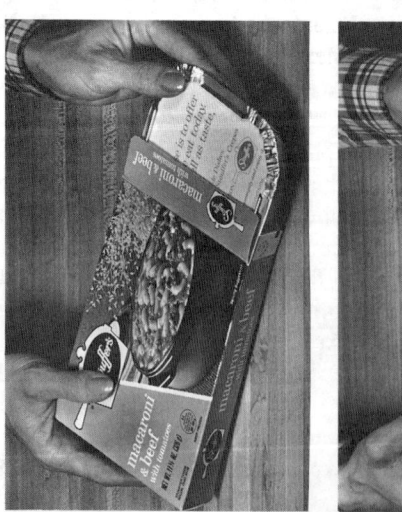
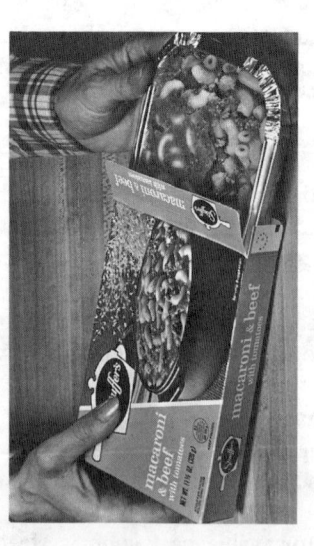

Fig. 4.5 Recommended procedure for heating TV-dinners in foil containers in the microwave oven. Step 1 (upper left): Remove the aluminum foil container from the carton. Step 2 (upper right): Remove cover from the container. Step 3 (lower left): Return aluminum foil container to the original carton, leaving one end-flap partially open. Cooking in the carton eliminates any risk of arcing. Step 4 (lower right): Place carton in microwave oven and heat. The carton will help in cooking and cleanup. Steam generated inside the carton helps food cook more completely and evenly, and any spillage resulting from liquid boiling over the edge of the tray will be absorbed by the carton rather than spilling in the oven.

(Courtesy of the Aluminum Foil Container Manufacturers Association) where it was originally developed by CMB Packaging S.A. of France. The can is plastic coated on the inside and outside: on the inside to prevent food contact with metal and outside to prevent metal contact with the oven wall. Even with the special plastic coating steel cans are cheaper than plastic cans and are 100% recyclable, a factor that is likely to become a popular selling point in the near future. The big problem, the steel people admit, is convincing the public that it is safe to place steel cans in the microwave oven.

### Making the package sell

The microwave package must announce and sell the product within. It should stand out among other packages. This it does by identifying the contents as microwavable, usually with a flag of some kind on the front of the package. There was an effort by the frozen food industry to provide a common, distinguishing mark, shield or other symbol such as used by other industries. It might be even wiser to adopt a symbol that is universal and not just for frozen foods. Some of the symbols that are being used today generally show a horizontal wavy line and the word "Microwavable."

Good graphics are being used by most food manufacturers to show off their products and these unquestionably influence consumer buying decisions. Some packages have all the graphics on the lid whereas on others the graphics are placed on a paperboard sleeve or the container is overboxed to provide protection as well as space for graphics.

### Packaging designed to control microwave heating

Very early in the microwave development effort, the pioneer in this field, Raytheon Company, became aware of the limitations of their brainchild and filed for patents on techniques to correct some of these limitations. A review of some of the early patent literature is both interesting and instructive.

## EARLY MICROWAVE PACKAGING PATENTS

Spencer (1949), probably as a consequence of frequent demonstrations of microwave corn popping, a rather showy performance without visible signs of heat, filed an application and was granted a patent that described the microwave popping of corn in plastic bags. The unique aspect of this patent was popping the kernels directly off the cob. The unpopped kernels remained on the cob for easy disposal (Fig. 4.6).

Spencer (1950) also conceived of a flexible film package or bag attached to a piece of stiff board with a fold out handle, similar to handles on paper coffee cups (Fig. 4.7). A tear string was provided to simplify opening the package of hot food. The handle made it easier to pour the contents out of the bag and lessened the risk of burns from escaping steam.

One of Spencer's co-workers and the first manager of Raytheon's Radarange Division, A.E. Welch (1950), realized the difficulty in heating a frozen dinner in a microwave oven. His solution to this problem was to package the dinner in a tray and wrap it in a stretchable film (rubber hydrochloride). Steam generated during heating was contained by the film and condensed on the cold food areas giving up its heat. This technique improved heating uniformity since the heat of condensation was given up wherever there were cold spots on the food surface. The film did not burst because of its stretchability and also because some steam tended to condense on the relatively cool film under the low ambient temperature conditions of the oven thus relieving the pressure (Fig. 4.8).

Moffet (1952) was granted a patent on the use of shielding so that the topping

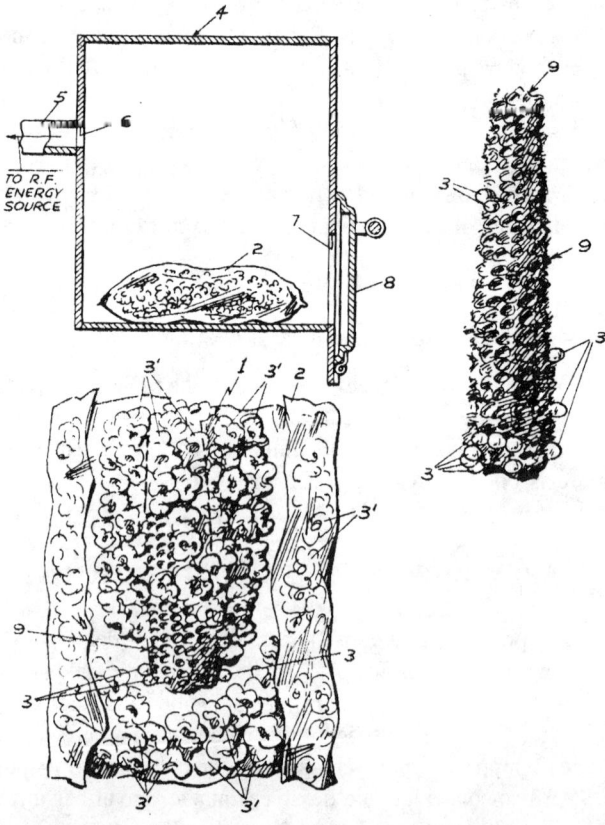

Fig. 4.6 Illustration of early patent for a method of popping corn in a microwave oven (Spencer, 1949).

of an ice cream sundae could be heated without melting the ice cream. Moffet pointed out that his invention could be applied to the heating of other packaged food products containing more than one food component where it is desirable to heat one item to a greater extent than another. He cited as an example, the heating of the filling of a sandwich.

Baker and Krajewski (1966) patented a technique and various packaging configurations to effect proper heating of a complete frozen dinner. One embodiment of their invention used a dielectric insert that fit into a metal tray such as a TV-dinner tray. The insert had compartments of different depths chosen to place the food components at different distances from the metal surface. Thus the energy passing through the food set up standing waves with the one quarter wave nodes at locations that insure maximum energy intensities in the foods. In this way the amount of energy absorbed by a particular food could be controlled by its distance from the reflecting metal surface. Another aspect of their invention was to provide a compartment that reduced the tendency for overheating at the periphery of foods by selectively increasing the distance from the reflecting surface to the center of the food (Fig. 4.9).

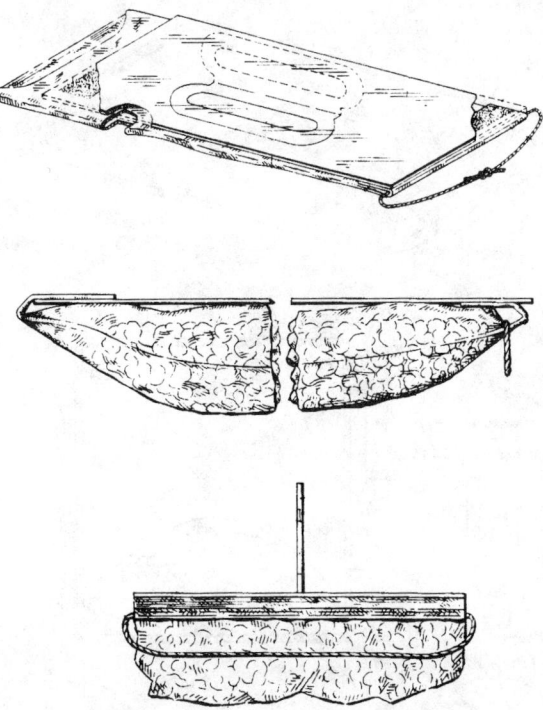

Fig. 4.7 Early patent for packaging food for easy and safe opening after microwave heating (Spencer, 1950).

## MORE RECENT PACKAGING PATENTS

Faller (1981) developed a package for heating food products that included integral support means and apertures for permitting moisture to escape from the package during microwave heating. These features are especially important when heating foods such as pizza and other foods that would tend to reabsorb moisture and give soggy, wet results. The practice of this invention accomplishes that objective while also elevating the product above the package surface where moisture would tend to condense and accumulate (Fig 4.10). A moisture barrier film is attached to tabs on the bottom surface so that when the film is removed the tabs will be displaced downward thus forming a support for the carton and also raising the product above the bottom of the carton. Apertures may be provided for venting moisture on the upper surface in the same manner.

A series of patents for novel packages and devices for popping corn in microwave ovens were granted to individuals associated with the Pillsbury Company. Gades

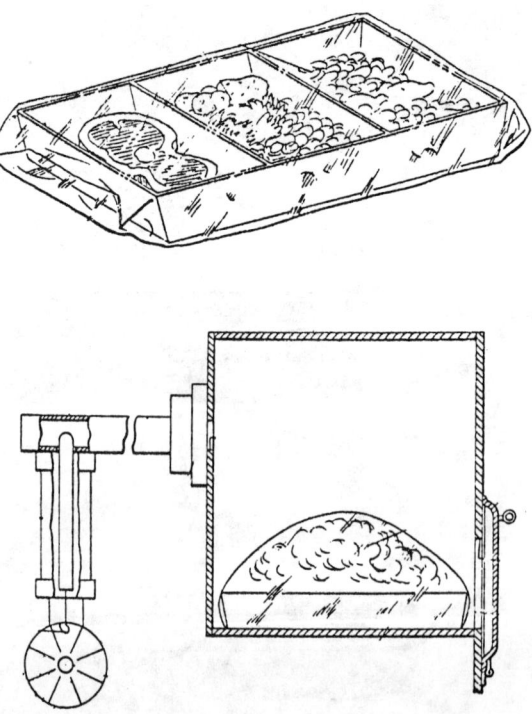

Fig. 4.8 Method for heating frozen meals (Welch, 1952).

PACKAGING FOOD PRODUCTS FOR THE MICROWAVE OVEN 103

et al (1974) were granted a patent for a device that effectively concentrates microwave energy to a specific location in a microwave oven. The device consists of a low dielectric loss material having two sheets of aluminum foil imbedded as concentric rings in a plane parallel to the oven floor and spaced about 3/4 to 1-inch below the product to be heated. It was found that when a package containing popcorn kernels was placed on the device a very substantial improvement in popping percentage occurred in the lower powered consumer microwave ovens. The main shortcoming of this device was its weight, about 5 pounds.

Brandberg and Andreas (1977) developed a popcorn package that combined a container to hold the charge of popping corn and a communicating expandable container to hold the popped kernels (Fig. 4.11). The problem that it is designed to overcome is the long time it takes to obtain substantial popping in consumer microwave ovens. During this time the popped corn is subject to scorching. The container holding the charge of popcorn ingredients is a self-supporting truncated cone shape where unpopped kernels will tend to fall back into the melted fat of the mixture where they have a better opportunity to pop. The interconnecting expand-

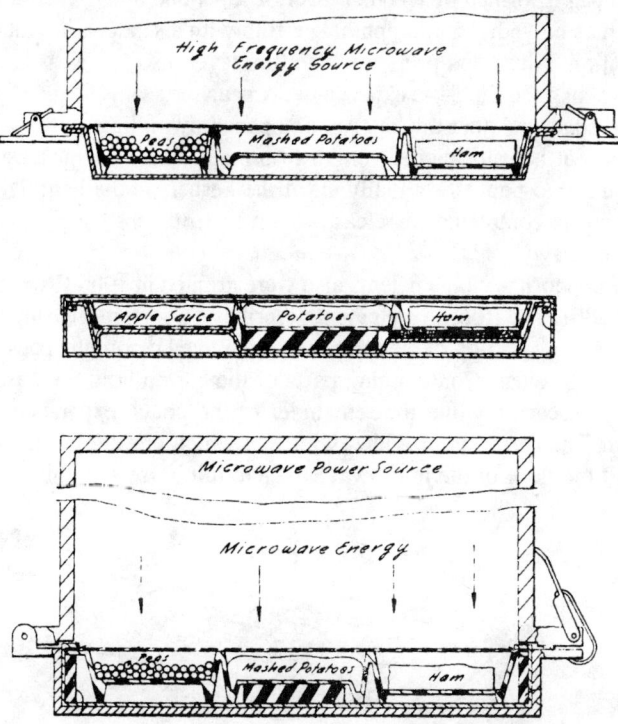

Fig. 4.9 Method to improve microwave heating of food
(Baker and Krajewski, 1966).

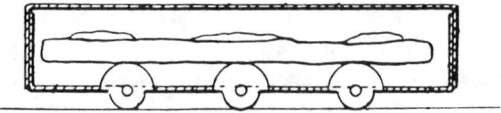

Fig. 4.10 Package for microwave heating
(Faller, 1981).

able container is provided with pleats and folds to enable it to increase many times in volume as it fills with popped kernels.

A patent granted to Borek (1980) demonstrated that more corn popped and greater volumetric expansion occurred when a piece of insulating material was attached to the package as shown in Fig. 4.12 By reducing heat loss to the oven floor in this way, better popping was achieved in the lower wattage ovens commonly found in the home.

Cage et al (1986) patented a shelf-stable combination of an easily openable bag and a mixture of edible popcorn ingredients suitable for use in microwave ovens. The gusseted bag is formed by an outer layer of paper and inner layer of non-wicking material such as polyethylene terephthalate film with a seal coating that is sensitive to a combination of heat and pressure. The film layer also serves as an oxygen barrier thus increasing the shelf-life expectancy. A pealable seal is formed by the coating along the top edge such that the bag can be opened by pulling on diagonal corners. The pealable seal has sufficient strength to remain closed for at least one-half of the time required to pop substantially all of the kernels in the bag. The seal opens before popping is completed to release steam and prevent the popped corn from becoming too chewy.

Two other popcorn package patents also were granted in 1986 (Roccaforte, 1986; Webinger, 1986). The Roccaforte patent describes a pouch in a carton arrangement wherein a tear away portion of the top panel is removed before the package is placed in the microwave oven. Constraining parts of the carton hold the basal portion of the pouch in the carton while the remainder of the pouch expands as it fills with popped corn. The carton is provided with feet which elevate the bottom panel of the carton off the floor of the microwave oven to minimize scorching of the bottom panel.

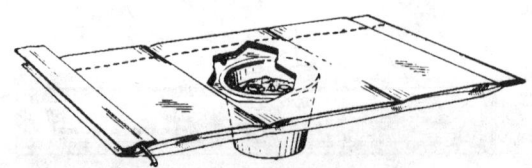

Fig. 4.11 Corn popping patent
(Brandberg and Andreas, 1977).

Fig. 4.12 Microwave popcorn package (Borek, 1980).

The Webinger patent also is a pouch in a carton arrangement including a truncated pyramid structure of paperboard that is adapted to be packaged inside the carton. When the package is to be used, the outer carton is inverted and opened, the stand is removed, erected and placed in the microwave oven. The inverted outer carton is then placed on the stand which effectively insulates the lower portion of the carton from the oven floor. In this position the popping area of the package is in a position to be affected by maximum microwave energy efficiency and is protected against heat loss to yield a higher ratio of popped to unpopped kernels.

Today, a number of companies are marketing popping corn in expandable popping bags and other novel containers for home microwave oven use. According to the *Federal Register*, December 6 (55 FR 50404), 1990, an individual has petitioned the FDA to allow use of poly (phenyleneterephthalamide) resins as a material from which reusable microwave popcorn popping bags could be made.

### Shielding

This is a microwave technique that takes advantage of the reflective property of metal, usually aluminum foil because it is flexible and can be wrapped around irregularly shaped foods to protect parts of foods from becoming overcooked. Early examples of shielding were the use of foil to wrap the tips of chicken wings and legs and the ends of rolled roasts for at least a portion of the cooking process. When this technique is applied to beef roasts it is possible to prepare a roast that is the same degree of doneness from one end to the other, an accomplishment that is practically impossible by any other cooking method at least not in typical cooking time periods. In addition, this method gives extremely high cooking yields.

**The differential heating container.** Shielding applications soon occurred to others and a number of patents were granted. One company (Teckton, Inc.) was formed to exploit this technology. Some rather novel designs were developed that are interesting to contemplate.

One problem that had plagued microwave and food technologists since the microwave oven was first introduced related to the difficulty in heating a complete dinner without overheating and damaging certain components of the dinner. A typical frozen dinner of meat, potatoes and a vegetable after heating in a microwave oven would find the vegetable overheated and dried out before the meat was at serving

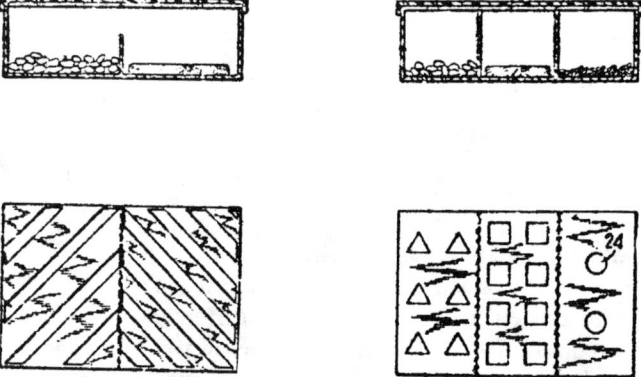

Fig. 4.13 Shielded frozen food package (Brown, 1965).

temperature. A number of patents were granted to correct this differential heating condition so that each component of a meal can be heated to its appropriate serving temperature. Brown (1965) recognized the difficulty in cooking two or more food portions that required different heating times and patented a technique for solving this problem. According to his patent the food portions are placed in individual compartments of an aluminum food container and each portion covered with foil having apertures of different sizes to permit different quantities of microwave energy to enter each compartment to heat the contents to the appropriate degree (Fig. 4.13).

An example of another approach (Stevenson, 1970) is shown in Fig. 4.14. Called a Differential Heating Container (DHC) by the firm (Teckton, Inc.) that was

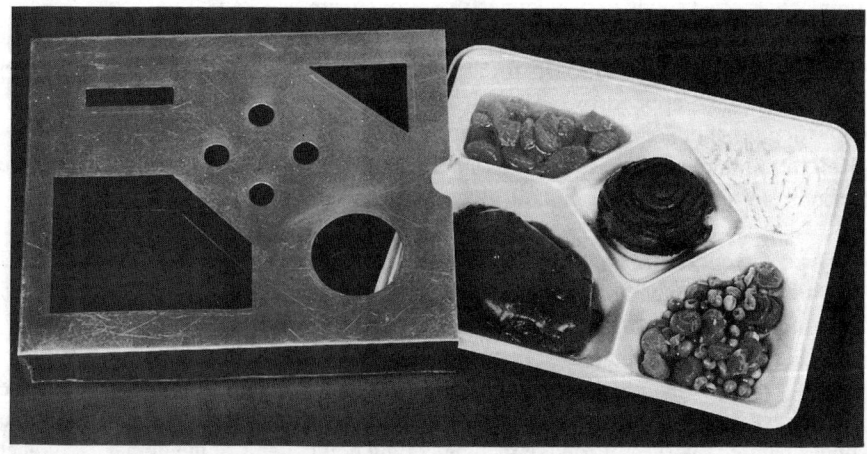

Fig. 4.14 Differential heating container (Courtesy of Teckton, Inc.).

founded to market the concept, this metal sleeve has similar sized openings in the upper and lower sides whose dimensions permit the passage of appropriate amounts of microwave energy to the food components when the dinner is placed in the sleeve. Thus the meat and gravy component, which because of its mass requires the most energy, is located between the largest openings, while the sweet roll, which requires only sufficient energy to become slightly warm, is placed beneath the small holes in the center; and the frozen orange sections, which need just a few calories to thaw, receive their portion of energy through the narrow slot in the upper left corner of the device. Variations of the DHC have been used in hospital foodservice, an elderly feeding program, a sandwich shop, and a convenience food store.

Fichtner (1967) preceded the DHC concept with a unique shielding technique illustrated in Fig. 4.15. It had long been known to microwave engineers that microwave energy will pass through wire mesh and that the amount of energy that would pass was related to the wire size and the distance between the wires. Fichtner's patent applied this knowledge to the controlled heating of meal components in a microwave oven. Thus the fastest heating components would be placed in the compartment with the smallest mesh size and the slowest heating foods in the compartment with no wire mesh at all.

In contrast to the Brown patent that teaches the use of metal containers with various sized openings on one side, and the DHC concept with openings on opposite sides, is the Micro-Match (Alcan International, Ltd) approach that claims to concentrate microwave energy in the package. The unique feature of the Micro-Match concept is the special lid that is meant to be used with standard aluminum foil containers (Fig. 4.16). The lid may be made from cellulosic or plastic resinous sheet or film having a low dielectric loss factor with imbedded foil segments or metal painted areas or other inclusions of metal. The plastic prevents the foil container from touching the microwave oven wall, hence no arcing can occur. But mainly, its function is to control the microwave heating of the contents, even selective heating of food in multi-cavity containers. In addition, it is claimed that its use can promote browning and crisping of the food surface. Infra red photography has shown much better heating

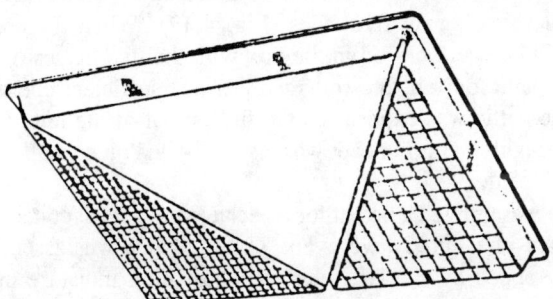

Fig. 4.15 Microwave cooking utensil (Fichtner, 1967).

Fig. 4.16 Microwave heating method using metal foil containers and plastic lid with foil panels (Courtesy of Alcan International, Ltd., Toronto, Canada).

results in metal containers using the Micro Match lid than with plastic or paper containers.

According to the applicable patent (Keefer, 1987), to paraphrase the claim, a food in a container, either metal or non-metal, capable of being heated in a microwave oven operating at 2450 MHz, having a non-reflecting cover with a dielectric constant greater than 10 spaced from the food surface a distance of 0.8 to 2 cm, so that the dielectric constant of the cover and the spacing of the cover above the food permit the passage of microwave energy through the cover into the package while interfering with reflected microwave energy within the package, thereby retaining and concentrating the microwave energy within the package.

A great deal of interest has been expressed in the Micro-Match concept by major food manufacturers, but only time will tell if this shielding approach fares better than its predecessors.

Pothier and Ford (1975) designed a package for heating sandwiches in which paperboard stock with aluminum foil laminated to an area that when folded up into a box provided partial shielding on five sides (Fig. 4.17). When the components of a hamburger sandwich were placed in the box with the two halves of the roll on the bottom and the pattie on top, the roll received just sufficient energy to thaw and warm, while the pattie was exposed to the full output of the microwave oven and was heated thoroughly. This concept was test marketed on a small scale but to date has not been adopted.

Woods (1977) was granted a patent for a package for a refrigerated or frozen sandwich to be heated in a microwave oven. The sandwich was assembled from two pieces of bread topped by a filler, overwrapped with a moisture impervious film and frozen. The bread base was at least partially shielded with a sheet of aluminum foil over 5 to 10% of its base area (Fig 4.18).

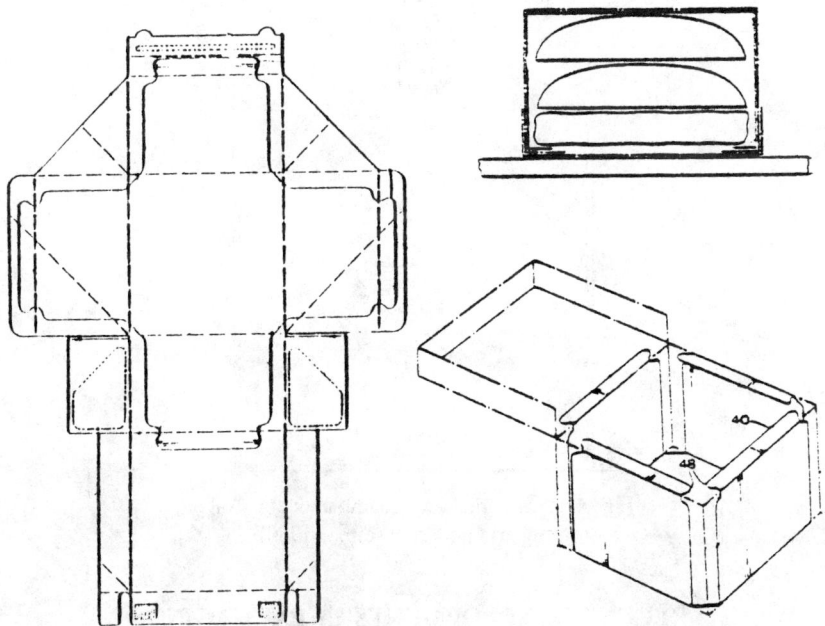

Fig. 4.17 Shielded package for microwave heating frozen hamburger sandwich (Pothier and Ford, 1975).

Slangan and Tsunashima (1980) were granted a patent for a novel ice cream sundae package in which syrup contained in a separate compartment above the ice cream is selectively heated when placed in a microwave oven while the ice cream remains frozen. This was accomplished by making the base of the syrup compartment opaque to the passage of microwave energy. After heating, the upper compartment base is punctured to permit the hot syrup to flow down over the ice cream. In another version of this package, the upper compartment was separable from the ice cream compartment and its contents after heating could be poured over the ice cream (Fig.4.19).

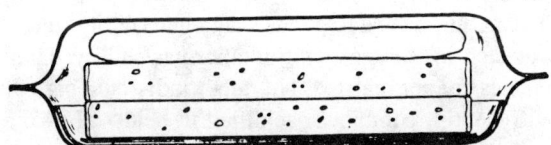

Fig. 4.18 Container for microwave heating frozen sandwiches (Woods, 1977).

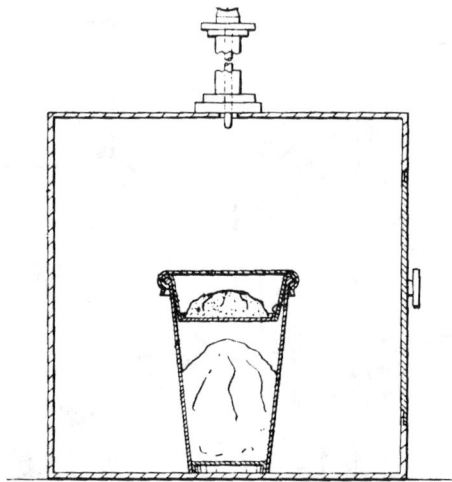

Fig. 4.19 Shielded ice cream sundae package (Slangan and Tsunashima, 1980).

## RECENT DEVELOPMENTS IN PACKAGING

The recent surge in new product introductions for the microwave oven market beginning about 1984 has been remarkable. Rice (1986b) commented on retortable trays that provide the necessary barrier properties for extended shelf stability. Plastic trays with EVAL (ethylene vinyl alcohol) or PVDC (polyvinylidene chloride) sandwiched between layers of polypropylene provide an adequate oxygen barrier and are suitable for microwave heating but cannot tolerate the high temperatures of conventional ovens.

A proprietary development by General Foods (Anon, 1986) is said to have extended the operating temperature range of plastic retort trays suitable for microwave or conventional oven heating to 375 F or better.

Glass packaging suddenly took on a aggressive posture and a number of glass-packaged pasta products appeared in national distribution after successful test marketing. Rice (1986c) reported on Cheseborough-Ponds introduction of its new Ragu™ Pasta Meals. The company chose glass because of its clarity (which allows customers to see what they are buying), its easy reclosability (thus enabling unused product to be protected in storage), and its microwavability. Unlike most canned foods, the Pasta meals are not retorted, but rather individual ingredients are cooked separately just sufficiently, combined, acidified to below pH 4.6 and hot filled at 195 F.

Microwave directions also can be found on glass packed vegetables, directing the lid be removed and a piece of plastic film be placed over the opening before heating.

There have been some unique developments in paperboard packaging that deserve mention. Kuchenbecker (1986) was granted a patent for a 2-piece microwave heating

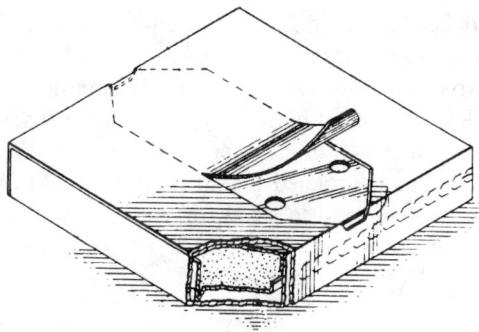

Fig. 4.20 Disposable container for microwave cooking (Kuchenbecker, 1986).

package with a removable section for exposing water vapor ventilation holes and an inner food supporting tray (Fig. 4.20). In one version the tray has V-shaped supporting legs that raise the food above the surface of the outer package and the surface of the tray has a microwave absorptive coating, which heats to brown and crisp the surface of the food in contact with the tray. In another version microwave reflective shielding material is applied to an inner top panel to prevent overcooking of the top of the food, as for example a pizza topping.

Brown and Maroszek (1986) patented a microwave cooking carton designed to brown and crisp foods on both sides, particularly foods having non-uniform dimensions like breaded chicken parts. This is accomplished by having a pair of opposed microwave interactive surfaces for converting microwave energy into conductive heat. Handles formed in the paperboard container permit the container to be inverted after the foods in contact with one surface have been crisped and browned to fall onto the other surface for finishing (Fig. 4.21).

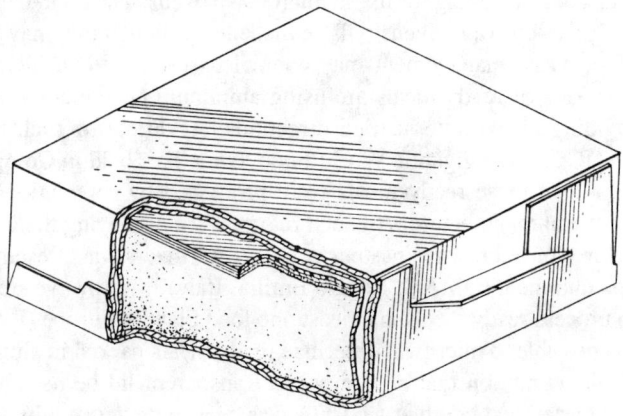

Fig. 4.21 Microwave cooking carton for browning and crisping food products on both sides (Brown and Maroszek, 1986).

A recently introduced breaded fish product comes with a pair of susceptor sheets, a coated paperboard tray and a foil lined paperboard sleeve with die cut elongated oval slots on the two narrow sides and one on the bottom side. To prepare the product, (1) the fish filets are placed in the tray between the susceptor sheets; (2) the tray is placed in the sleeve then in the microwave oven; (3) the assembly is heated on high power for 5 minutes, rotated 1/2 turn and heated 3 to 4.5 minutes longer. The assembly should be allowed to stand for a minute before removing from the oven, presumably because it will be quite hot. Interestingly, there are some cautions noted on the package:

Do not place on metal rack or on any metal surface in microwave.
Do not use if package is damaged or torn.
Do not allow package to touch sides of microwave.
Do not microwave more than one package at a time.
Do not reuse package.

## PACKAGING AND THE ENVIRONMENT

The Financial Times, October 23, 1991 (Gooding, 1991) carried a special section on aluminum the gist of which was that this metal has a promising future because: 1) it is recyclable and substantial tonnage of aluminum beverage cans are being recycled, the quantities will continue to increase, and other aluminum packaging materials are being added to the recycling program; 2) the automotive industry will begin to use very substantial quantities; and 3) it is environmentally friendly.

It is very likely in view of environmental concerns that larger quantities of aluminum will be used for food packaging for use in microwave ovens. The myth that aluminum cannot be used in microwave ovens will be dispelled. Though this may be difficult the problems of waste management may compel its renewed consideration.

European packers of ready meals are using aluminum containers with ring pull tops and providing microwave heating directions. A variety of such meals were displayed at ANUGA, the gigantic World Food Trade Fair held in Cologne August 12–17, 1991. All of these ready meals were processed by conventional retorting methods and several manufacturers of such retorts were displaying their equipment. One manufacturer of microwave pasteurizers and sterilizers was present to discuss its processing equipment (OMAC, Reggio Emilia, Italy). Microwave sterilizers are being used to process ready meals in at least one food plant in Europe (P & T Foods, Belgium). It is possible to microwave sterilize ready meals packed in aluminum containers under the condition that a microwave transparent lid be used.

The Alcan International booth at ANUGA was backed up with equipment for collecting and crushing foil containers to be recycled into aluminum ingots for ultimate production of new foil containers. The message that foil containers can be used in microwave ovens was supported by literature. Indirect support was provided by other literature being distributed at ANUGA and in trade publications about Der Grüne Punkt (The Green Point in English) that in Germany means recyclability and reusability are the law. The Green Point is a symbol to mark that a package is recyclable or made from recycled materials and it involves a financial contribution to the German Dual System of collection and utilization of packaging waste. These contributions are to be used to finance a nationwide collection and sorting system for packaging materials that can be recycled. According to an article in the ANUGA '91 magazine handed out at the Fair, by the 4th quarter of 1991 some four billion packages will carry the Green Point. The intent of the Dual System is to replace the present environmentally unsound disposal system with an ecologically sound recycling system. It is expected that implementation of this system will relieve the mountains of waste and incineration plants of no less than seven to eight million tons. Yellow trash cans or bags are being distributed throughout cities for recyclable packaging that will be picked up by special disposal companies. The contents will then be sorted and forwarded to the responsible packaging industries for reprocessing. Opinion polls show that most consumers in Germany are willing to pay a few cents more for reusable packaging if it will have a beneficial impact on the environment.

In the United States, at Pack Alimentaire '91, both Alcoa and Reynolds announced that they were accepting aluminum formed containers in their recycling programs in addition to beverage cans.

In the U.S., Luigino's in Duluth, has introduced 3-lb family-sized aluminum pans of frozen Italian foods such as Fettuccine Alfredo. The pan is coated and has a paperboard lid that extends outward so that the pan cannot touch the oven walls, thus eliminating the possibility of arcing. The coated pan makes it esthetically pleasing to go from oven to table. Heating time of 20 minutes in a microwave oven is seen as acceptable to consumers.

## SUMMARY

"Packaging remains the key for total acceptance of the microwave oven and the convenience food approach to foodservice at the consumer and commercial level" (Anon, 1978). This quotation says volumes. The package protects, sells, may function as the serving container, and controls and assures successful microwave heating performance.

Packaging developments are coming fast and furious. Co-extrusion technology promises to provide the solution to microwave retort pouches that today require an aluminum foil layer to insure adequate gas barrier properties. Manufacturers are aiming at the market for hot fill, sterilizability and retortability. Asceptic packaging was expected to reach 3 billion packages by 1990 (Anon, 1983). Dramatic changes in packaging methods and materials are being brought about by consumer preferences and changing technology. Much of this demand can be attributed to the growth of the microwave oven market.

But not all food products need be developed for in-package heating. The microwave food market will be much broader even though the convenience aspect of in-package heating will tend to outpace the less convenient preparation of food from scratch. The cake and cookie market still require reusable microwave utensils, and some rather unique designs to improve cooking results can be expected.

As an indication of the commitment that some food companies realize must be made to exploit the microwave phenomenon, Frank H. Terwilliger, Director of Packaging for the Campbell Soup Company commented (Anon, 1984): the Campbell Soup Company is committed "to providing microwavability in their line of convenience food products." Terwilliger added that plastic containers are a top choice for their products. The market has directed that packaging must provide convenience and microwavability and at least one year of shelf-life. A "thrust toward a more upscale market" was indicated with an effort to supply "an elegant product packaged in an outstanding way—one that will be table-ready and totally pleasing to the senses . . ." In view of environmental concerns about disposable packaging some rethinking of how this can be accomplished will be necessary. Truly, the packaging industry has a formidable challenge facing it.

## REFERENCES

Anonymous (1978). Convenience foods and associated microwave ovens. Frost & Sullivan, Inc., 106 Fulton St., New York, NY 10038.

Anonymous (1984). Food industry forum. Food Processing 45 (4), 72.

Baker, B.J. and Krajewski, E.Z. (1966). Food package for microwave heating. U.S. Patent 3,271,169.

Borek, J.R. (1980). Microwave popcorn package. U.S. Patent 4,219,573.

Brandberg, L.C. and Andreas, D.W. (1977). Combined popping and shipping package for popcorn. U.S. Patent 4,038,425.

Brandt, Y.M. (1979). Packaging and utensil perspective: the shape of things to come. Microwave Energy Appl. Newsl. 12 (4) 9,10,13.

Brown, E. (1965). Frozen food package and method for producing same. U.S. Patent 3,219,460.

Brown, R.K. and Maroszek, R.V. (1986). Microwave cooking carton for browning and crisping food on two sides. U.S. Patent 4,590,349.

Cage, J.K., Kolla, S. and Gates, J.E. (1986). Container and popcorn ingredient for microwave use. U.S. Patent 4,571,337.

Colato, A.E. (1978). Microwave properties of materials for microwave cookery. Microwave Energy Appl. Newsl. 11 (6) 3-6, 13.

Decareau, R.V. (1977). The effect of aluminum packaging materials on microwave oven performance. Final Report, The Aluminum Association, Washington, DC.

Faller, R.A. (1981). Food carton for microwave heating. U.S. Patent 4,260,060.

Fichtner, E.C. (1967). Microwave cooking utensil. U.S. Patent 3,302,632.

Gades, L.D., Brandberg, L.C. and Gorman, R.A. (1974). Composite microwave energy perturbing device. U.S. Patent 3,835,280.

Gerling, J.E. (1981). Research confirms suitability of conventional glass containers for use in microwave ovens. Food & Drug Packaging, January 8, 20–22.
Gooding, K. (1991). The metal of the future. Financial Times, Oct. 23, p. 29.
Hecht, J. and Haskell, J.V. (1976). New high barrier films which permit microwave heating. In *Abstracts of Presentations*, Microwave Power Symposium, Leuven, Belgium.
Keefer, R.M. (1987). Microwave heating package and method. U.S. Patent 4,656,325.
Kuchenbecker, M.W. (1986). Two-blank disposable container for microwave food cooking. U.S. Patent 4,592,914.
Laperle, E.A. (1988). High barrier plastic food containers—the organoleptic challenge. Proc. ESL '88, Extended Shelf-Life for Foods Conference, Orlando, Florida, February 9–10.
Lyons, T. (1988). Status report: PVDC resin for rigid barrier food containers. Presented at ESL '88, Extended Shelf-Life for Foods Conference, Orlando, Florida, February 9–10.
Moffet, W.F. Jr. (1952). Method of heating frozen food packages. U.S. Patent 2,600,566.
Monte, W.C. and Landau-West, D. (1983). Expanded polystyrene containers in microwave cookery. J. Am. Dietet. Assoc. *83* (3) 323–327.
Pothier, R.G. and Ford, T.E. (1975). Partially shielded food package for dielectric heating. U.S. Patent 3,865,301.
Rice, J. (1984). Coated aluminum tray with protection dome tops. Food Proc. *45* (9) 112–113.
Rice, J. (1986a). GF breaks new ground in convenience packaging. Food Processing *47* (12), 21–22.
Rice, J. (1986b). Contract packaging in retortable plastic trays. Food Proc. *47* (4) 160–161.
Rice, J. (1986c). Reinvigorated marketing thrust for glass-packed foods. Food Proc. *47* (8) 106–107.
Risch, S.J. (1988). Microwaveable packaging materials: regulatory concerns. Proc. Pack. Alim. '88, March 22–24, San Francisco.
Roccaforte, H.I. (1986). Microwave popcorn package. U.S. Patent 4,584,202.
Slangan, G. and Tsunashima, M. (1980). Ice cream package including compartment for heating syrup. U.S. Patent 4,233,325.
Snedeker, R.H. and McKenna, L.A. (1978). A comparison of microwave cookware materials. In *Digest, Microwave Power Symposium 1978*, p. 107–109, Ottawa, Canada.
Spencer, P.L. (1949). Prepared food article and method of preparing. U.S. Patent 2,480,679.
Spencer, P.L. (1950). Receptacle. U.S. Patent 2,528,251.
Stehle, A.P. (1979). Packaging and utensil perspective: shape of things to come. Microwave Energy Appl. Newsl. *12* (4) 13–15.
Stevenson, P.N. (1970). Selective cooking apparatus. U.S. Patent 3,615,713.
Webinger, G.P. (1986). Food Package. U.S. Patent 4,586,649.
Welch, A.E. (1950). Method of treating foodstuffs. U.S. Patent 2,495,435.
Woods, F.J. (1977). Container for the microwave heating of frozen sandwiches. U.S. Patent 4,015,085.

# CHAPTER FIVE

# BROWNING AND THE MICROWAVE OVEN

## INTRODUCTION

We eat with our eyes, as the saying goes, and although foods come in many bright colors we tend to judge many foods by their various shades of brown. Breads and pastries must be golden brown; steaks and chops, burnt or charred; and toast is not toast unless it is brown and crisp. How mundane eating would be without the browning reaction. Against this background of consumers being accustomed to food browned to a turn come gentle microwaves with all the browning ability of a hot water bottle.

The ambient temperature in a microwave oven rarely reaches much above room temperature. Because of this and the penetrating ability of microwave energy, evaporative cooling occurs at the surface of foods cooked by microwaves and the result is a negative temperature gradient; that is, a higher temperature inside the food than at the surface. Thus the two factors responsible for browning to occur are essentially absent: time and temperature. Browning either does not occur at all or, with few exceptions, is inadequate.

Spencer (1952) probably recognized this deficiency of microwaves before the first microwave oven came off the assembly line and filed a patent application for a technique that he believed provided a solution. His solution involved the application of materials to the surface of foods that would absorb microwave energy at a greater rate than the food and thus create a toasted surface. The examples cited by Spencer were: pastry dough baked until hard, then ground into fine particles that could be sprinkled on the pastry items to be baked, or made into a paste with water and spread on the food item: and meat cooked to a charred dehydrated state, similarly ground and applied to the surface of meat products. Other examples cited by Spencer included the use of chestnut flour, egg mixtures, and other substances containing organic oils and salts. One might say that Spencer was an unsuspecting advocate of the edible susceptor.

## BROWNING FORMULATIONS

Copson, Neumann and Brody (1955) working at the Raytheon Company Food Laboratory studied the carbonyl-amine reaction, the reaction first proposed by the French chemist Maillard in 1912, on the development of surface browning during microwave cooking. Their thesis was that if the compounds responsible for the browning reaction were applied in greater concentration at the surface of foods, the reaction might be triggered at a lower temperature and therefore provide characteristic browning in microwave cooking time.

Formulations of glucose and other sugars, and glycine, a simple amino acid, in various ratios, were evaluated. Small amounts of sodium carbonate were added to adjust the pH of the reaction. A mixture of 40% glycine, 40% xylose and 20% sodium carbonate to give a pH of 9 began to color at 160 F and gave a charred appearance at 172 F. Two grams of this formula sprinkled on a 59 gram pork chop gave good color in 1.25 minutes at 0.75 kW of microwave power.

## Browning formulations for meat

Copson et al obtained browning of roasting chickens by applying a mixture of 25% salt, 25% non-fat dry milk solids, 20% glucose, 20% glycine and 10% sodium carbonate blended with egg white. The egg white improved the adherence of the mixture to the chicken and gave more uniform browning.

Almost 20 years passed before further research was reported on microwave browning. Baldwin and Brandon (1973) compared a number of browning formulations in browning chicken breasts and pork chops. Among the materials tested were the xylose, glycine, sodium carbonate formula mentioned by Copson et al; a brown gravy mix, Worcestershire sauce, and Kitchen Bouquet. The treated foods were cooked in microwave ovens and presented to panelists for their judgments of browning and uniformity of color, flavor, and general acceptability. The brown gravy mix was judged most successful and the xylose mixture least successful in browning chicken breasts and pork chops. The other two treatments were rated satisfactory for pork chops.

Microwave oven cookbooks also list browning agents and techniques for treating food surfaces. Methven (1978) cites a number of commercial steak sauces, barbecue sauces, soy sauces and the like that can be brushed on meats before microwave cooking. Chicken parts can be brushed with melted butter and sprinkled with paprika, onion soup or gravy mix. Bouillon granules can be sprinkled on meats. The effect of all these treatments is quite satisfactory in enhancing the appearance of foods.

Some formulations such as Micro-Shake (product of Golden Dipt Corp.) have salt as a major ingredient, that when applied to the wetted surface of meat increases the electrical conductivity of the surface. Electrical conductors absorb microwave energy avidly and produce higher temperatures at the surface. The result is good browning. This technique also reduces microwave energy penetration so that much of the internal cooking is by heat conduction that in the case of thin meat patties does not add much to the total cooking time. The overall result is better yield, less of the juices are cooked out of the pattie and an attractive natural appearance is obtained.

Moody (1981) patented a formulation for microwave browning that consists of an aqueous syrup comprising a caramelized foamed disaccharide, either alone or in combination with a minor amount of a monosaccharide, typically 0.01 to 0.1 parts by volume to reduce a slight bitterness sometimes detected in syrups made from disaccharides alone. Microwave cooked foods that ordinarily come from the oven with a white or grayish appearance can be basted with this syrup, or the syrup may be included in a recipe to obtain a desireable browned appearance.

## Dough browning formulations

Copson et al obtained good coloring of pie crust when 3.5 grams of a formulation of 40% glycine, 40% glucose and 20% sodium carbonate was added to 284 grams of pie crust mix. Even more satisfying results were obtained when some of the formulation was added to a wash of milk and egg yolk then applied to the top crust. These additives were described also as having enhanced the taste of the products in which or on which they were used.

Fulde and Kwis (1984) patented a reactive dough mixture that resulted in surface browning when exposed to microwave energy. The composition consists of a dough base with added reducing sugar and an amino acid source to promote surface browning in microwave time. Pie crust for frozen meat pies, for the purpose of this invention, is one containing approximately 50% shortening. Conventional baking of this dough for 45 minutes at 425 F will give a golden brown color, but the dough will not brown in microwave time. To effect microwave browning, yeast extract should be added to the dough at least at the 0.5% solids level, and the concentration of yeast extract should be in the 7 to 13% range. Reaction time appears to be related to the level of yeast extract and the presence of moisture. At low levels of extract, browning will be completed in a sealed product container in about 8 minutes. At intermediate levels, browning will occur in a non-air tight product container, while at high levels, browning will occur without the need for a cover.

The preferred method according to the patent is to apply, uniformly to the surface of a substrate pie crust that has been brought to 50 to 60 F, a coating of one of the following compositions:

TABLE 1.
BROWNING FORMULATIONS

| Ingredients | Formula A | Formula B (Parts by weight) | Formula C |
|---|---|---|---|
| Pie dough flour | 38.0 | 38.0 | 38.0 |
| Shortening (Colfax) | 24.5 | 23.7 | 18.5 |
| Water | 31.0 | 30.2 | 27.9 |
| Salt | 0.8 | 0.8 | 0.8 |
| Dextrose (fine) | 1.3 | 1.3 | 1.3 |
| Yeast extract (70% Solids) | 4.4 | 6.0 | 13.5 |

Adapted from Fulde and Kwis (1984)

Fellenz and Moppett (1991) working for the Pfizer Corporation developed a novel browning formulation that can be sprayed onto products such as raw pie crusts prior to packaging and freezing. When the product is microwave heated the browning reaction is triggered and a full brown color develops. The heating process results in full color development. The formulation is said not to impart flavor or odor notes.

## BROWNING DEVICES

The early history of microwave browning devices was centered around two developments: the so-called browning dish and ferrite technology. Both of these technologies have become commercial realities and represent significant sales dollars in the market place. Both are based on the conversion of microwave energy into heat in a supporting utensil and the transfer of that heat to the surfaces of foods at temperatures sufficient to brown.

### The browning dish

The microwave browning dish originated, serendipidously it seems, in the laboratories of the Corning Glass Works. In the 1930s Corning scientists were investigating electrically conductive coatings, particularly those based on tin oxide (Panzarino, 1975). These studies led to patents being issued (Mochel, 1951) and the production of electrical resistors and space heating panels. Warming trays, percolators and a dish that heated when placed in a microwave oven were produced in the early 1950s.

The browning dishes being marketed today (Fig. 5.1) are made from a glass-ceramic substrate with a tin oxide coating on the underside. The dishes are similar to Corningware but slightly modified in shape and provided with feet to prevent contact with the microwave oven floor so that no heat is lost to the floor by conduction and the maximum thermal energy is applied to the food. The tin oxide coating interacts so efficiently with the microwave energy that almost complete energy-to-heat conversion occurs. About 4 to 6 minutes in a microwave oven are required to reach temperatures of 500 to 600 F. Care in handling the browning dish is essential as severe burns are possible by accidental contact with the skin. These dishes also may be used for general microwave oven purposes without preheating.

### Ferrite browning devices

The application of ferrite technology to food browning originated in the Raytheon Company research laboratories as an outgrowth of more esoteric applications. Ferrites are ferri- or ferro-magnetic compounds such as barium and strontium titanate that change from magnetic to non-magnetic at a specific temperature referred to as the Curie temperature. The Curie temperature is typically above 500 F, but can be

Fig. 5.1 Microwave browning dish (Courtesy of Corning Glass Works, Corning, New York).

adjusted by formulation. These materials readily absorb microwave energy and convert it to heat. When the ferrite reaches the Curie temperature it becomes nonmagnetic and essentially microwave transparent. As it cools it regains its magnetic properties and will begin to heat again.

Copson and Davis (1958) described a number of ferrite utensils and devices for browning, searing and crusting foods in microwave ovens. When the Curie temperature is reached there is a reduction in the process of converting microwave energy into heat and the energy will pass through the device to heat the food product in contact with it. Additional ferrite browning patents include: Derby (1976), Freedman (1976), McMaster and Dudley (1976a,b) and Teich and Dudley (1984).

The application of this technology to grilling meats in a microwave oven is the basis of a steak grill currently on the market (Freedman and Bowen, 1982). The design of the grill represents an improvement over earlier technology in that ferrite members are not in direct contact with the food item, but rather transfer heat to metal members that are in direct contact. Thus cooking (searing) is accomplished by conduction with heat created indirectly in the metal (See Fig. 1.31 and 1.32).

## Susceptors

Browning dish technology has been expanded recently to include substrates other than glass or glass-ceramics. A patent assigned to the Pillsbury Company (Winters et al, 1981) describes the use of salt hydrates such as calcium and lithium bromides, calcium and lithium chlorides and magnesium chloride, either singly or in combinations, with a polar solvent such as water. The patent refers to these materials as "chemical susceptors," a susceptor being a device for converting microwave energy into heat that in turn heats another material placed against it. Success requires the susceptor to heat faster than the food article to be heated. Continued heating of the material occurs until a certain maximum temperature has been reached. Thereafter, the susceptor becomes transparent to microwave energy. The result is similar to ferrite behavior except that magnetism does not appear to be involved (Fig. 5.2).

An objective of this invention is to provide an inexpensive, flexible and disposable package for crisping and browning foods such as pizza, french fried potatoes and the like. The combination of solute (susceptor) and a polar solvent (water) should depress the vapor pressure of the solvent by 25%. More highly soluble solutes that will depress the vapor pressure by 70 or 80% are preferred. Most of the inorganic salts mentioned in the patent will form pastes with water and can be applied to suitable substrates.

An unusual approach to browning foods in a microwave oven is described in patents assigned to General Mills (Brastad and Beall, 1980; Brastad 1981). These patents teach the use of metallized film to convert microwave energy into heat that then is transmitted directly to the food surface to brown and crisp. Intimate contact of this film and food are essential to success because of the low heat capacity of the film (Fig. 5.3). The patent specifically mentions thin coatings of aluminum applied

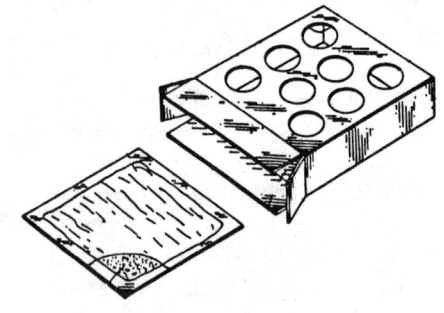

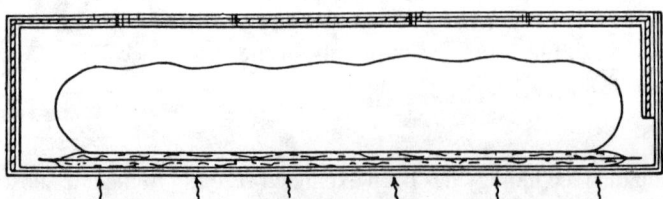

Fig. 5.2 Package incorporating chemical susceptors for browning products such as pizza (Winters et al, 1981).

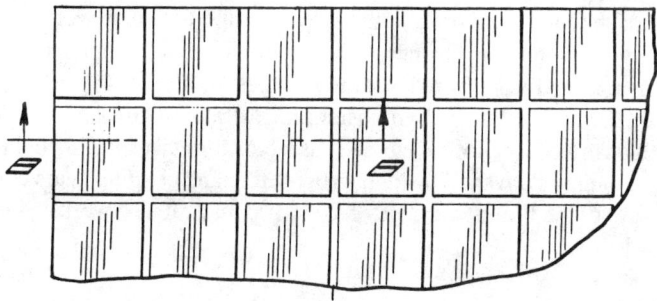

Fig. 5.3 Metallized film for microwave browning of food products (Brastad and Beall, 1980).

by vacuum evaporation on a polyester substrate. Metal coatings with a thickness of 0.1 microns or less will permit microwave energy to be transmitted to a considerable degree. The metal coating preferably should have a resistivity of 1 to 10 Ohms/sq. in. The coating is divided into a myriad of metal islands separated by narrow dielectric gaps formed by means of suitable screens or masks during vacuum evaporation. The islands, in effect, become the plates of capacitors causing high voltages to develop that cause pronounced heating to occur. A retail product in roll stock was announced by an Australian firm, Leigh-Mardon.

Brastad pointed out that the film can be supported by more rigid materials such as paperboard. A number of food products are using these susceptors, as they are called, for browning and crisping. A major supplier of susceptor board is James River Corporation. Products include pizza, French fries, and breaded fish portions. Some popcorn packages also incorporate susceptors to improve popping efficiency.

Andreasen (1988) speaking on susceptors in microwave packaging pointed out that for browning and crisping susceptors must be able to heat to 300 to 400 F rapidly and maintain those temperatures for 5 to 10 minutes. Because of the requirement not to exceed the flashpoint of packaging materials, around 450 F, Optical Coating Laboratory, Inc. chose to employ sputtered 316 stainless steel in its susceptor development efforts. Stainless steel has its maximum absorption (50%) at a relatively low resistance, while reflectance and transmission of the metallized film are around 25% each. Thicker layers result in lower microwave absorption, less transmission and greater reflectance as would be expected. Conversely, thinner layers have greater transmission and less reflectance and absorption. The package designer thus has the versatility to provide sensible heat (by absorption) to the food product by susceptor means while shielding it from a significant part of the energy or allowing the food to be exposed to a greater degree of microwave heating. Conceivably, this approach could be applied to a multi-component dinner as well as to a single food component.

Although sputtered aluminum currently is dominant in susceptor packaging, stainless steel has an advantage in that it is stable to oxidation and thus maintains its ability to heat in microwave ovens for longer periods of time. Andreasen (1988) presented data showing that with stainless steel, temperatures between 200 and 420 F, obtained by varying the resistance, can be maintained for as long as 30 minutes. One firm (Printpack, Inc.) offered a stainless steel metallized polyester film at the 26th Annual Microwave Power Symposium in Buffalo in August 1991. The same firm also demonstrated a cellophane based susceptor film that shrinks slightly around a product, then vents clam shell fashion as browning and crisping nears completion. A suggested use of the film is with flaky bakery products. It also works with a variety of sandwich items.

## Oven browning means

A few countertop microwave ovens are equipped with resistance heating elements similar to those used in broilers, though usually of lower power. This was the approach to browning used in the first consumer ovens built by the Tappan Company in 1956. Microwave ovens designed to operate at 115 volts cannot operate the browning element simultaneously with microwave energy generation. Typically, partial microwave heating is carried out first, then the browning element is energized to provide surface color as well as some additional cooking.

Combination microwave and conventional ovens, either as in free-standing ranges or built-in wall ovens, are made to function as a conventional oven, a microwave oven or with both energy sources applied simultaneously. Usually, in the combination mode the microwave power is operated at a lower level. Proper adjustment of microwave and thermal energy gives good cooking results with natural browning.

Microwave forced convection ovens were introduced as a commercial item in the late 1960s by Hirst (Microwave Heating) Ltd. under the name ARCTICAIR, and later in the United States as the Micro-Aire oven. This oven has thermostatically controlled hot air heated by resistance elements up to 550 F and variable microwave power up to 2.25 kW. One of the more dramatic applications of this oven is on British Rail in galleys for prime cooking of meats, fish and poultry. An interesting sidelight is the almost exclusive use in the oven of metal skillets and other shallow metal cookware. A consumer version of the forced convection microwave oven was introduced by several manufacturers in the United States.

## Microwave frying

It is possible to fry foods in deep fat in a microwave oven, though no specific devices have appeared on the consumer market for this purpose. Frying is accomplished by heating a quantity of cooking oil in a suitable container in the microwave oven to the appropriate frying temperature, adding a quantity of food and continuing to cook with microwave energy and the hot fat until the food is properly fried. This was demonstrated in the early 1950s at Raytheon Company's Food Research Laboratory in a conventional frying tank built into a commercial microwave oven. The cooking oil was heated by conventional means and microwave energy was applied to the product in a wire basket immersed just below the fat surface as illustrated in Fig. 5.4 (Pierce and Copson, 1961).

Some degree of caution must be exercized as with any frying method. Glass or glass-ceramic containers are not recommended because of the possibility of breakage and spilling the hot fat. Shallow, heavy aluminum containers with a diameter of 9 inches or more will permit efficient heating of the cooking fat. The container should have handles made of non-conducting materials that can tolerate high temperatures.

Fat and oil have a low specific heat and will heat about twice as fast as water. Unlike water, however, they will continue to heat to the smoke point and beyond, thus it is necessary to monitor the temperature of the fat or oil during heating.

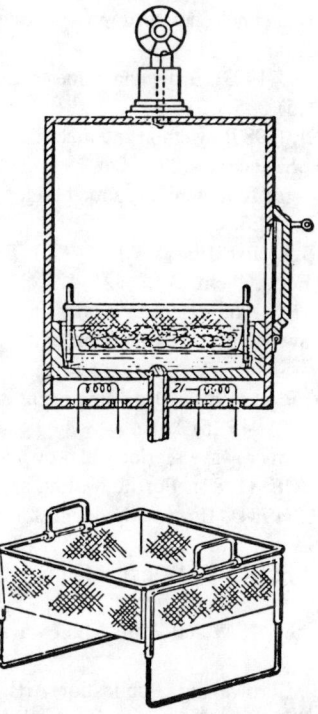

Fig. 5.4 Device for combined microwave and hot oil frying of food products (Pierce and Copson, 1961).

Microwave ovens with built-in temperature probes are particularly suitable for this purpose, especially those that can be set to turn the oven power off when a certain temperature has been reached.

## SUMMARY

Microwave browning technology is far from being exhausted and new unique products and methods can be expected to appear from time to time. Only recently, susceptor technology was applied to vending of microwave heated frozen french fries. Several hot air and microwave devices were introduced in 1991 for vending french fries and pizza. Browning of singular components of frozen meals is within the realm of possibility.

## REFERENCES

Andreasen, M. (1988). New technologies to improve susceptor efficiencies in microwave packaging. Proc. Pack Alim. '87.

Baldwin, R.E. and Brandon, M. (1973). Browning of meats cooked by microwaves. Microwave Energy Appl. Newsl. 6 (5) 3-5.

Brastad, W.A. and Beall, N.J. (1980). Method and material for prepackaging foods to achieve microwave browning. U.S. Patent 4,230,924.

Brastad, W.A. (1981). Packaged food item and a method for achieving microwave browning thereof. U.S. Patent 4,267,420.

Copson, D.A., Neumann, B.R. and Brody, A.L. (1955). Browning methods in microwave cooking. J. Agric. and Food Chem. 3 (5) 424-427.

Copson, D.A. and Davis, L., Jr. (1958). Heating method and apparatus. U.S. Patent 2,830,162.

Derby, P.P. (1976). Microwave heating apparatus with browning feature. U.S. Patent 3,946,188.

Fellenz, D.C. and Moppett, F.K. (1991). Browning agent enhances visual appeal of microwaved foods. Food Technol. 45 (6) 111.

Freedman, G. (1976). Folding microwave searing and browning means. U.S. Patent 3,949,184.

Freedman, G. and Bowen, R.F. (1982). Ferrite heating apparatus. U.S. Patent 4,362,917.

Fulde, R.C. and Kwis, S.H. (1984). Brownable dough for microwave cooking U.S. Patent 4,448,791.

MacMaster, G.H. and Dudley, K.W. (1976a). Microwave browning means. U.S. Patent 3,934,106.

MacMaster, G.W. and Dudley, K.W. (1976b). Microwave browning utensil. U.S. Patent 3,946,187.

Methven, B. (1978). Basic Microwaving. Publication Arts, Inc., Minneapolis, Minnesota.

Mochel, J.M. (1951). Electrically conductive coating on glass and other ceramic bodies. U.S. Patent 2,564,566.

Moody, R.D. (1981). Microwave cooking browning composition. U.S. Patent 4,252,832.

Panzarino, J.N. (1975). Development of the microwave browning dish. Appliance, June.
Pierce, B.C. and Copson, D.A. (1961). Microwave apparatus. U.S. Patent 2,997,566.
Spencer, P.L. (1952). Electronic cooking. U.S. Patent 3,582,174.
Teich, W.W. and Dudley, K.W. (1984). Microwave heating method and apparatus. U.S. Patent 4,454,403.
Winters, W.C., Chang, H.-H., Anderson, G.R., Easter, R.A. and Sholl, J.J. (1981). Microwave heating package, method and susceptor composition. U.S. Patent 4,283,427.

# CHAPTER SIX

# NEW PRODUCT DEVELOPMENT FOR THE CONSUMER MICROWAVE OVEN MARKET

By the early 1970s at least one food manufacturer was marketing plated diets for hospital food service. These frozen dinners would later find their way into the supermarkets and would be joined by others as it became apparent that there was a good market among microwave oven owners for high quality frozen dinners. All of these plated dinners would be tailored for dual oven capability (conventional and microwave) as a hedge against the possibility that the microwave oven might be a passing fad, but they also recognized that a very large conventional oven market existed and probably would exist for many years to come. A long time would pass while food manufacturers awaited the development of an ideal dual ovenable container. As pointed out in the chapter on packaging, the aluminum container filled this roll, but remained unacknowledged for years.

It is interesting to reflect in some detail on certain of these efforts, particularly on some of the food products. But before doing so, let us examine a list of twenty suggestions for product development for the microwave oven market cited more than a decade ago to put the subject into better perspective (Decareau, 1975). Some of these may seem obvious, and indeed there is considerable activity today in a number of these cases. Others may provide some incentive to food manufacturers.

1. French fried potatoes are clearly the most popular of frozen foods, but easily one of the most difficult challenges facing the microwave food product developer. A solution to this problem could be a bonanza to the developer. There is a huge market potential in both hot food vending and fast food operations, and an even larger consumer market. In addition, a product with a lower fat content is a distinct possibility.

2. The hamburger sandwich. The usual result out of a microwave oven is a hot, soggy, tough roll and a slightly warm piece of meat. A large hot food vending market and an even larger fast food market awaits a solution to this problem.

3. Hot submarine sandwiches. A novel approach to a microwave oven version of this popular product was test marketed more than 15 years ago. The problem of heating the filling without overheating the roll was solved by the use of a unique shielding device. As shown in Fig. 6.1, the filling is placed in one compartment of a polyurethane foam container and the opened roll in a parallel compartment.

Fig. 6.1 Microwave shielding device (Courtesy of Teckton, Inc., Framingham, MA).

The container is overwrapped with film or lidded with a peelable film and frozen. To heat, the packaged submarine sandwich is placed in the shielding device, which provides maximum exposure for the filling, and then into the microwave oven. After heating, the film is removed and the section of the container with the roll folded over onto the section with the filling and inverted, placing the filling in the roll. A variety of such sandwiches was tested in a foodservice operation in 1975.

4. New sandwich concepts. A wide variety of typical and novel sandwiches for microwave oven preparation is possible. Bars, cocktail lounges, vending and the consumer market represent the potential market.

5. Pizza pies. A solution to give a crisp crust typical of these products might be provided by the use of specially designed microwave browning dishes.

6. Complete meals. Pick and choose meal combinations from a variety of individually packaged frozen or thermally processed components. A great opportunity for functional packaging exists here. The containers could be disposable or reusable. Possible applications include vending machines in apartment houses, college dormitories, military barracks, and other locations unable to justify conventional foodservice. Meal components to meet the whims of children is one variation.

7. Do your own thing foods. Consider some possible basic entrees which permit the homemaker to exercise her own creativity. For example, a basic beef stew, could be converted into a goulash, a beef Burgundy, an Irish stew, a curry, etc. The same is possible with other meats, fish and poultry.

8. Gourmet specialties for those special dinner parties. Many of these items are available, though not in the supermarket. It is not necessary that all items be prepared by microwave methods alone. The homemaker is not adverse to following direc-

tions in which the microwave oven is used only for a part of the preparation.

9. Family-size or individual vegetable dishes with uniquely fresh flavor. The package could be the heating and serving dish. Individual servings of corn-on-the-cob are an excellent example and could find a ready customer in vending foodservice.

10. Family-size or individual pot pies. The pie crust should be packaged separately to be heated in either a hot oven or on a microwave browning dish. A tie-in with a reusable microwave transparent container as a premium might enhance sales.

11. Delicatessen take-out foods. Consider a do-it-yourself heating of Deli items at a self service counter in a supermarket. The option of heating the item at home in a microwave oven is always possible. A variation on the standard containers now used in delicatessens should not be too difficult to develop.

12. Salads and desserts. Consider frozen packaged salads and desserts that require only a brief microwave exposure before serving.

13. Ice cream sundaes. A package design is needed which permits the topping to be heated or thawed without melting the ice cream. The use of shielding techniques would appear to offer a solution to this problem.

14. Pasta dishes. A large variety of precooked pasta dishes (such as spaghetti in its many variations, lasagna, manicotti and others) would find a ready market for home consumption, vending and specialty fast food operations.

15. Other ethnic dishes. Chinese foods, Mexican foods, Swedish meat balls, coq au vin, chicken paprika, fish and chips, and many more all should find ready acceptance.

16. Breakfast foods. Many foods, which have found only marginal acceptance because of limited quality, are much better when microwave heated. Examples include French toast, pancakes in many variations, omelets of all kinds and ready to eat cereals.

17. Crepes. Many variations of this popular food are excellent when microwave heated. A popular restaurant chain is a large user of microwave ovens for heating crepes. It should be possible to make crepes for the consumer microwave oven market.

18. Soups. Use your imagination here. The possibilities are almost unlimited. Packages which function as the serving bowl and are disposable should not be too difficult to develop. The homemaker should welcome such items which allow her to serve a variety of soups without the need to wash any sauce pans.

19. Pastries. Pies, Danish pastries, donuts and many other possibilities suggest themselves. For certain items, the microwave browning dish offers a potential answer.

20. Frozen meats. At least one airline uses, or did use, pre-seared steaks and chops in the First Class section. Raw frozen meats have been test marketed in the past, but have not been very successful against fresh meats. With today's high labor costs, the economics should be more favorable, and microwave tempering offers the means for the supermarket meat department to provide the shopper with the option of purchasing frozen or tempered meat cuts.

## PRODUCT DEVELOPMENT ACTIVITY

A review of some of the literature follows to set the stage for a logical and scientific approach to microwavable food product development activity. Some of the literature relates to the previous list of categories.

### Cake mixes

Copson (1957) reported on a successful procedure for microwave baking of Angel food cakes. In this work, carried out at the Raytheon Company Food Research laboratory, the effect of beating time, container, microwave power level and auxiliary heat were studied. Quality comparable to conventional baking was obtained when a paper container was used, under beating of the batter compared to that provided by package directions, and a short application of top heat from resistance heaters near the end of the baking cycle. Baking time was 4.5 minutes at 800 watts of microwave power with auxiliary heat during the last 2.5 minutes. A commercial prepared mix was used.

A layer cake can be baked in a few minutes in a microwave oven, and for this reason there is much interest in developing a product for the microwave oven. Generally layer cakes have been less than satisfactory. In the first place, they do not develop a brown crust. This however can be masked by suitable toppings. There is also a tendency for wet spots to remain on top, which can only be eliminated by some overbaking.

Martin and Tsen (1981) noted that microwave baked cakes had slightly better volume if the water level of the batter was increased to 115%. Batter flow that occurred during microwave baking could have been caused by the volume increase at the periphery. Reducing the heating rate at the edges by redesign of the baking dish should be attempted before any concerted reformulation effort; however, promotion of a special formulation for microwave baking can be a marketing advantage.

The Procter and Gamble Company introduced a cake mix under the Duncan Hines label in 1982 especially for the microwave oven. The objective was to provide a more satisfying microwave baked cake (Anon, 1982). This was accomplished, it was claimed, by adding special shortening and leavening agents. New ingredients in this line of cake mix that are not found in the company's regular mixes include wheat starch, corn starch, lactostearin and guar gum. The latter two ingredients are purported to give a somewhat smoother texture and batter. Monocalcium phosphate, one of the elements of the leavening composition reduces the amount of carbon dioxide during mixing by 67%. The cake mix is convenient to prepare. Just add egg, water, oil; mix and bake in the same mixing bowl for ten minutes.

The Pillsbury Company introduced a packaged microwavable brownie mix that includes a reusable polycarbonate baking pan. The pan is rectangular but with generous radii in the corners. Refill packages can be purchased for making additional batches of brownies using the same baking pan. The directions specify addition of ¼ cup of oil and ⅓ cup of hot tap water; mixing well; spreading evenly in the pan and baking about 4 minutes at high power (600 watts), turning the pan after 2 minutes. The batter does rise first around the edges during the first half of the baking cycle, but then the central area catches up to give a relatively uniform finished product. The result is a tasty, slightly chewy brownie.

## Breaded products

These have always been a difficult problem because of the movement of moisture out of the product and into the breading, tending to make it soggy and rubbery. A crisp product seemed unattainable until a recent announcement by National Starch and Chemical Corporation. Predipping fish fillets in a solution of waxy maize starch followed by drying prior to applying batter, parfrying and freezing is said to give a crisper product when microwave cooked. It is claimed that the starch forms a film on the fish that helps to seal in moisture and to prevent it from soaking into the outer coating.

A microwavable breaded shrimp that can be cooked from the raw frozen state in a microwave oven was introduced by Treasure Isle, Inc., about 1984. The key to the success of this product was a moisture resistant breading and a batter, both of which were identified as trade secrets. The breading was based on an oriental-style of breading imported from Japan that resembled rice kernels in appearance. The product was sold in a 16-oz package with 14-oz of shrimp and two 1-oz packets of dry sauce mix (one mustard, the other sweet and sour).

The effect of microwave cooking on the texture of battered and breaded fish products was studied by personnel at the College of Fisheries, University of Washington, Seattle (Lopez-Gavito and Pigott, 1983). They found that when the conventional batter formulation was modified by substituting partially hydrogenated soybean oil for water and using a modified waxy maize starch in the batter slurry, the texture was highly improved. Indeed, the sensory scores for the microwave product were similar to those for the deep frozen product. They found also that heating at a reduced power level (327 Watts) was less critical than heating at high power.

Of the five batter ingredients tested (two modified corn starches, a waxy maize starch, yellow corn flour and soft-white wheat flour), the waxy-maize starch had the lowest water absorption. Of the four breadings (pregelatinized barley flour, yellow corn meal, modified corn starch and double toasted Japanese style breading) the double toasted breading absorbed the least water vapor.

Subsequently six products were prepared for evaluation: three from minced fish patties and three from whole fish fillets. Two of each were battered using 47% modified corn starch, 52% water and 1% salt; and one of each was battered using 44% waxy-maize starch and 56% partially hydrogenated soybean oil. All were breaded with double toasted breading. One of each with conventional batter was deep fried for 1 minute at 350 F. The remainder were microwave heated at 327 Watts for 1 minute.

Generally, deep fat fried products that had not been frozen had higher mean scores. Those with the modified formulation scored higher than those with the conventional batter when both were microwaved. Of the samples reheated after cooking and freezing, those with the modified batter and breading when microwaved scored as high as the deep fat fried product. Preference rankings for the heated frozen samples were highest for the microwave battered and breaded with the modified formulation followed by deep fat fried samples and the microwave conventional battered samples. Interestingly, products with the modified batter formulation contained considerably less fat than deep fried products.

**French fried potatoes**

This product also has not yielded easily to efforts to produce a crisp product out of the microwave oven. A patent was issued in which a crinkle cut approach (Fig. 6.2) was claimed to enhance crispness (Saunders and McLaughlin, 1980). Gorfein et al (1978) analyzed the chemical and structural changes that occur in potato tissue during processing. The extensive studies carried out included, in addition to evaluating different retail and institutional products such as straight and crinkle cut potatoes of different thicknesses: frozen extruded-type potatoes, fresh potatoes cut into different shapes and sizes, combinations of frying and drying with coatings. The range of coatings included: starch, alginate, carboxy methyl cellulose, hydroxy propyl methyl cellulose, methyl cellulose, calcium chloride solution, whole milk and non-fat milk solids.

All of the retail and institutional products gave typical rubbery, soggy results when microwave heated. Partial drying and frying for two minutes gave a frozen product that on reheating was significantly crisper. Moisture content was 40.5% and fat 14.3% compared to commercial frozen french fried potatoes that after reheating had 51.7% moisture and 11.7% fat.

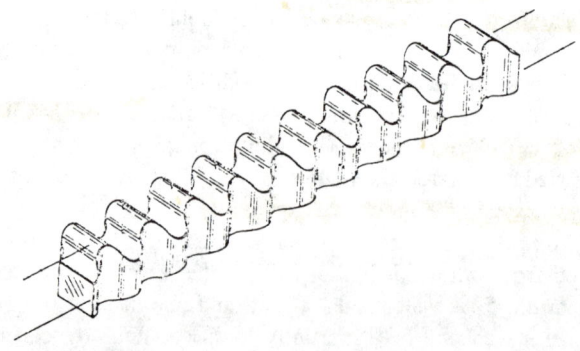

Fig. 6.2 Microwavable french fried potato (Saunders and McLaughlin, 1980).

In 1984 a frozen french fried potato product was introduced to test market by the Simplot Company, Caldwell, Idaho. Said to have been eight years in development, the product, sold under the label "MicroMagic," is a natural, crinkle cut potato that heats golden brown on the outside in less than two minutes for a 3-oz serving. According to a trade publication (Anon, 1985), "MicroMagic" fries had a phenomenal success in test market and the product is now available nationally.

A patent assigned to the Simplot Company described a process that produced a high quality microwavable french fried potato (Pinegar, 1986). The process comprises frying potato strips under controlled time and temperature conditions to include two parfrying steps with an intermediate cooling step wherein the strips are preferably frozen to reduce the moisture content of the strips in accordance with a predetermined relationship of strip size, surface area and moisture loss during processing. An empirically determined constant, called the Surface Area Constant (SAC) that provides an indication of quality characteristics, is the product of the average exposed strip surface area (PSA), the average strip percentage moisture (PML) and the average moisture loss in grams (GML). When the SAC value is $0.39 \pm 0.07$ the microwave reheated product is crispy with a moist mealy interior and rated superb. Analysis indicated 43% water and 14.8% fat. Other fried potato products also rated highly when the SAC was in the range indicated.

Other products on the market use susceptor boards in an effort to provide a crisp product. Good product-to-susceptor contact is essential if good results are to be obtained. It is not unusual to find that the susceptor boards have warped so that there is relatively poor contact. At least one product marketed through vending machines used an array of U-shaped susceptors (Fig. 6.3) and an extruded potato formulation. The product is contacted by susceptor surfaces on three sides.

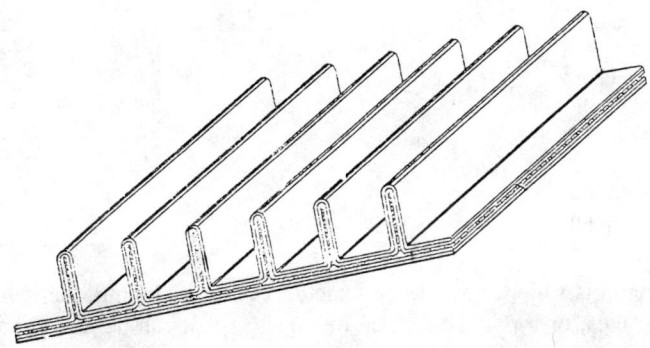

Fig. 6.3 Susceptor device for microwavable french fried potato product (Andreas and Cox, 1990).

## Pancakes

Norris et al (1976) recognized some of the problems associated with heating precooked frozen pancakes and other similar foods in microwave ovens that needed to be solved in order to provide a product suitable for microwave heating. The objectives as described in their patent were:

(a) to find a way to provide a package of precooked pancakes or other farinaceous griddle food without having to apply syrup or other sugar containing topping as a separate operation after heating;

(b) to provide a stack of pancakes or the like ready for heating in a container;

(c) to provide a stack of pancakes ready for heating with syrup applied;

(d) with provision for causing the pancakes to absorb microwave energy faster and more uniformly than plain pancakes;

(e) to provide a pancake with the syrup or sugar based topping that stays in place after being applied and before heating;

(f) to provide syrup topped pancakes in which the syrup is distributed so as to help absorb microwave energy over a wide area;

(g) with provision for preventing the absorption of odors and flavors during storage;

(h) with provision for preventing the absorption of syrup by the packaging material; and

(i) with provision of a foamed syrup topping for stacked farinaceous foods.

The major problem of providing a sugar based topping that would remain in place during storage and distribution was solved by the use of gelling agents in the formulation. A suitable formula from the patent is given in Table 6.1.

TABLE 6.1
MAPLE FLAVORED SYRUP

|  | % by weight |
|---|---|
| Margarine | 30.0 |
| Unflavored gelatin | 0.5 |
| Carboxymethylcellulose | 0.2 |
| Water | 9.0 |
| Maple flavor | 0.5 |
| Powdered sucrose | 12.0 |
| Corn syrup (liquid) (11.5% water, 36.3% solids) | 47.8 |

The dispersion is whipped to a foam structure. After whipping, the foam is spread on the pancakes, or it may be formed into a sheet that can be placed on each pancake, or it may be applied as a series of spaced apart dots (the preferred method if automatic equipment is used, since it can be applied in an extruded plastic condition). The dots will coalesce on heating. Such topping will remain in place without being absorbed by the pancakes, yet when heated will melt and trickle down over the pancakes to give the appearance of a freshly prepared product with a buttered syrup poured over them. To contain this product, a semi-rigid paperboard carton will suffice with top, bottom and side panels joined to the bottom with tucks and folds to prevent leakage of the syrup upon heating. A film overwrap is added to prevent absorption of flavors or odors. A polyester laminate on the inner surface is desirable to prevent absorption of hot syrup by the paperboard (Fig. 6.4).

The product is distributed frozen, but is usually dispensed from a vending machine or otherwise provided to the consumer for heating in a microwave oven. A heating time of 30 seconds in a 650 watt microwave oven is usually adequate. Typically, 72 grams of product (three pancakes) and 60 grams of topping make one serving. After heating, the container can be opened to provide its own serving dish. Results indicate that most of the energy is absorbed by the syrup and gives more uniform heating results by it being spread over and through the stack. The quality of the syrup topping does suffer from handling in the distribution chain.

## Soups

Hot soup, always popular as a first course or even as a meal, is a simple microwave item and should receive much greater attention in foodservice than heretofore given. In the early 1950s, Raytheon Company designed a microwave oven for snack bars and other foodservice operations. The oven shown in Fig. 6.5 represented the result of a cooperative effort between Raytheon and a major manufacturer of canned soups to develop a system to heat individual portions of soups quickly on customer request. Since this involved simply emptying a portion-size can of soup into a microwave compatible serving dish and heating, success should have been assured. Oven costs and the poor maintenance posture of microwave ovens in those early days resulted in the soup manufacturer withdrawing its support.

Excellent quality soups have been produced then chilled by the KAPCOLD process (W.R. Grace & Co.) and delivered to foodservice outlets in special CRY-O-VAC bulk (two gallon) packages. They are heated at the outlet, typically, in a

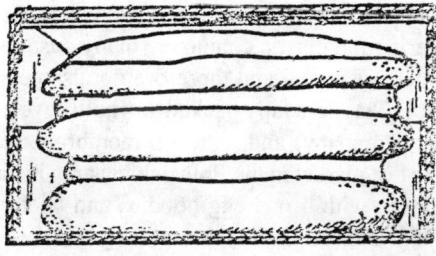

Fig. 6.4 Microwavable pancake product (Norris, 1976).

138    MICROWAVE FOODS: NEW PRODUCT DEVELOPMENT

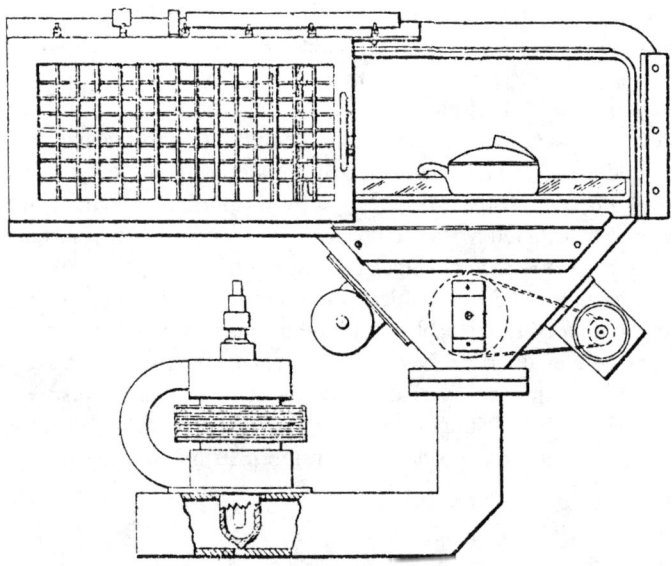

Fig. 6.5 Microwave oven designed for heating soups in a small restaurant operation (Haagensen, 1958).

trunion kettle, but could be portion packed and heated as individual servings in vending machines or other foodservice operations. The asceptic pack currently used for juices could be used with some modification to permit use of a spoon. Shelf-stable soups in a retortable, coextruded plastic bowl have been test-marketed by the Campbell Soup Company (Fig. 6.6). Heat'n Eat™ Chicken Noodle soup in microwavable plastic bowls were introduced in late 1983 to a test market in the San Francisco area. This was followed in 1985 by Swanson Cup-O-Broth™ in chicken and beef flavors packed in American Can Company retortable plastic Omni™ containers. These were lidded with pull-top metal lids, which left a metal rim on the container when it was removed. Market results appeared to indicate customer resistance to using them in microwave ovens because of the metal rim even though tests have shown that arcing will occur only if the rim contacts the oven wall. Campbell's also found that with the label covering the plastic container, many customers seeing the metal lid assumed the container was metal and therefore not usable in microwave ovens.

Campbell's Cookbook Classic soups packed in multi-layer, co-extruded plastic bowls (Continental Can Company) and with foil membrane lids (Reynolds Metals Company) introduced to test market in the Philadelphia area included: vegetable beef, chicken vegetable, chicken with broad egg noodles and beef vegetable with broad egg noodles. These 9 1/4 ounce bowls of soup are packed in an over-carton with open end windows. Instructions tell the user to remove the foil lid and cover the bowl with plastic film when microwave heating.

PRODUCT DEVELOPMENT FOR THE CONSUMER MICROWAVE MARKET 139

Fig. 6.6 Microwavable soup product (Courtesy of the Campbell Soup Company, Camden, NJ).

Campbell's has also test marketed their Soup du Jour line of frozen soups in plastic (co-polymer of polypropylene) microwavable bowls. The line includes: clam chowder, cream of asparagus, broccoli, and spinach, and onion soup. Heat seal film closures are left on during microwave heating to prevent spattering. In addition, Campbell Soup Company has tested American Bounty™ chunk style soup with meat in microwavable plastic packages.

Other firms also have entered the microwavable soup market. Tuscon Dairy Farms, Inc., Union, New Jersey, came in with a line of frozen soups under the RSVP label. Soups include: tomato Florentine, New England clam chowder, green pea with ham, cauliflower au gratin and cream of broccoli. These 7-ounce servings take 2½ to 3 minutes at high power to heat to serving temperature. Myers Foods, a division of Hanover Brands introduced 10 varieties in July 1985. They were bean and ham, chicken corn, chicken noodle, cream of broccoli, cream of mushroom, cream of potato, New England clam chowder, seafood bisque, split pea and ham and vegetable soup. A high density polyethylene container was used.

## Snacks

The Pillsbury Company introduced microwave popcorn under its Hungry Jack label in 1976, initially for sale through vending machines. The popcorn is a mix of coconut oil, seasonings and corn kernels in an expandable, laminated, duplex, greaseproof package (Fergusson, 1976). It can be stored frozen for six months; however, the instructions call for holding at 40 F at least 24 hours before loading into vending machines. Two minutes in a commercial microwave oven (1000 watts) gives ½ gallon of popped corn (Fig. 6.7).

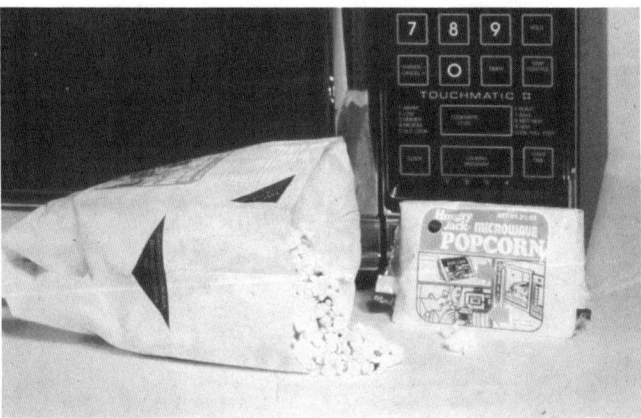

Fig. 6.7 Microwavable popcorn

Today there are probably a dozen or more suppliers of microwave popcorn on the market with products varying from plain to caramel coated, cheese flavored, and light salt to name a few. One microwave popcorn product comes with a packet of cheese flavoring to be added to the popped corn. In the ten years between 1980 and 1990 the microwavable popcorn market grew from $32 million to $898 million. This represents over 7% of all snacks sold.

It should be possible to develop a broad variety of other kinds of puffable snacks, though probably without the pyrotechnics of popcorn. The key to puffing appears to be creation of some degree of surface or case hardening of the material so that steam can be generated inside by microwave heating to cause puffing to occur. Various flavors can be provided as post puffing coatings as with the various flavored potato, corn chips and corn curls.

Van Hulle et al (1983) developed food compositions and methods for preparing sugary coated puffed snack food products by microwave heating. The food compositions comprise a plurality of puffable farinaceous dough pieces or pellets and a puffing medium throughout which the pellets are dispersed. The puffing medium comprises from 40% to 95% by weight of a nutritive carbohydrate sweetening agent and about 5% to 10% moisture. The water activity of the food compositions is less than 0.75. To make a coated snack, a gelatinized dough is formed and shaped into pieces that are partially dried then dispersed throughout the puffing medium and microwave heated to puff the pieces and enrobe them with the fluidized medium.

Mickle et al (1980) were granted a patent for a method of manufacturing a high protein snack food. One of the objectives as implied in the title is the production of a more nutritious snack food. Typically crisp snacks such as potato chips, corn chips and the like are high in sugars and starch and low in protein. This patent describes a product made from whey, a by-product of cheese making to give a snack food with 2 to 3 times more protein and ½ to ¼ as much sugar or starch.

Koshida et al (1982) described a process for making dry fruit chips from apples, peaches, melons, apricots, persimmons and papayas. Two to 6 mm thick slices are freeze-dried first, followed by microwave drying under vacuum (10 to 30 Torr) to thaw and evaporate some moisture and evenly disperse the water soluble sugar concentration throughout to 10 to 40%, followed by vacuum drying to reduce moisture to less than 5% by weight. Water soluble sugars may be taken from the classes: mono-, di-, tri-, tetra-, poly-, and sugar alcohols. Sucrose is preferred. Freeze-dried fruits have been evaluated as inclusions with dry cereals some years ago and also as dessert items in military rations. Such products lacked the quality of good mouth feel, though flavor was rated highly.

More than 15 years ago "half-snacks" had been demonstrated by National Starch & Chemical Company as essentially pre-forms for snack foods. They came in several categories: extruded and dried; sheeted, scored and dried; and semi-moist. The latter could be oven heated to expand and finish. Presumably, these "pre-forms," "half-snacks" could be microwave expanded. Flavorings and spices were incorporated in the matrix before heating. The dried items were fried to produce snacks.

There are also snacks, ready made, to be warmed in the microwave oven. According to Anon (1985), consumers indicated a preference for warm snacks as confirmed by a DuPont Company sponsored focus group test. Potato chips, corn chips, popcorn, cheese curls, peanuts and a bridge mix in packages lidded with DuPont's Mylar OL (ovenable lidding) film were heated in microwave ovens for the test.

McCormick & Co.'s microwavable Nachos — tortilla chips with a cheddar cheese — dip were introduced to test market under the Tio Sancho label in July 1986. They are probably in national distribution at this time. The chips are in a metallized foil bag and the dip is asceptically packaged using the Conofast thermoform/fill/seal system of Continental Can Company. A high impact polystyrene tray is provide in which to heat the snacks. A 30 second microwave cycle melts the dip and warms the chips.

Real Fresh, Inc., which packages the McCormick cheese dips, introduced two asceptically packaged dips under their own Muy Fresca label. Asceptic packaging provides at least a year's refrigerated shelf-life.

Considering the success with microwave popcorn, it is difficult to imagine the challenge not being taken up to develop other microwave snack food products.

## The hamburger sandwich

The main problem in heating sandwiches, particularly frozen sandwiches, is excessive heating of the bread component in order to heat the filling to an acceptable temperature. In the usual sandwich configuration microwave energy must pass through the bread in order to heat the filling. Since the moisture content of the bread is low (ca 35%), relatively little energy is required to raise its temperature to an acceptable level, yet a longer heating cycle is necessary to insure that the filling is heated. This results in the bread being overheated, some of its moisture boiled off, and if the sandwich is wrapped much of the moisture will condense and be reabsorbed by the bread, making it soggy. If the sandwich is not wrapped, the lost moisture will result in a tough bread component and as the bread cools it also will harden rapidly.

The solution would appear to be a simple one: heat the bread and filling separately and let the customer assemble the sandwich. A second approach is that cited in Chapter Four; the patent of Pothier and Ford (1975) provides a means to heat the filling and bread together without excessive heating of the bread. Briefly, the package controls the heating by partial shielding of the bread component.

From a pure quality point of view, controlled heating of the bread component, rather than an attempt at reformulation of the dough would appear to offer the most likely avenue to success. Bread formulas that are purported to resist moisture loss may have been an improvement, but did not succeed because the heating rate was not controlled. Spreading butter on the roll or providing some other barrier to moisture loss was only partly successful.

Formulations high in sugar, shortening and egg and low in moisture are claimed to absorb microwave energy slower than standard bread and roll formulations. Toasting the bread to reduce the moisture content has been tried. Even the use of day-old bread has been suggested. None of these approaches have given particularly satisfactory results.

The patent of Ottenberg (1985) claims that a yeast-raised, wheat flour based formulation, including a percentage of rice starch, gives a bread product that has improved resistance to deterioration in palatability when microwave heated. The basis of this claim is the smaller crystal size of rice starch and certain other starches having a crystal size of 20 microns or less and constituting 10 to 20% of the total flour of the formula. The formulations given in the patent also cite the addition of wheat gluten to make up for the reduced amount of flour. Shortening, egg and syrup or sugar are also included in the formulation. When made into Kaiser rolls and microwave reheated to 130 to 140 F, taste panelists found the patent formulated product retained its palatability while the traditional Kaiser rolls were unpalatable because of a tough crust. No examples were given of sandwich heating, but it was stated that food products made from this formulation: e.g., pizza, sandwiches, bread and rolls should maintain their palatability.

Microwave heating of refrigerated sandwiches is very much less of a problem than heating frozen sandwiches. From a storage and distribution point of view, the frozen sandwich is a more desirable approach, particularly for the development of a successful consumer product. It is in this area that serious research is needed.

## Other sandwich items

Edge shielding with strips of aluminum foil or foil laminated to paper or paperboard could be used to protect the edges of sandwiches from the toughening effect of excessive heating. Shielding has been successfully used in heating frozen submarine sandwiches. Hot sandwich plates, such as meat and gravy on bread, could be provided in a number of ways. A package with a built-in shield could protect the bread component while optimizing the heating of the meat component. The sauce or gravy could be heated simultaneously in a separate container and poured over the sandwich by the customer.

There would appear to be substantial opportunities for developing a broad line of frozen sandwiches and considering the popularity of sandwiches such a development would be well worth the effort.

## Pizza pies

Crisp crust pizza is being provided now by the use of susceptor boards. However, the possibility of government restrictions on current metallized film susceptors because of the migration of adhesives or other materials into the food products in contact with them suggests that it would be wise to consider other ways in which the desired condition could be obtained. Some patents claim that crispness can be provided in other ways.

Bone and Manoski (1981), for example, teach that the use of a baked cracker-type dough with a moisture content of 5% or less, underlying and in contact with a baked bread dough with a moisture content in the 20 to 40% range (Fig. 6.8), will yield a very palatable pizza after cooking in a microwave oven. With the bread dough on top the cracker dough crust portion absorbs excess moisture during microwave cooking of the pizza.

Stangroom (1976) teaches that a pizza crust baked and dried to 3 to 8% will be crisp when microwave heated from the frozen or non-frozen state. He claims that a low shortening content and predrying the crust are keys to the success of this invention. Further, a low gluten flour is preferred. He also claims that tortilla dough

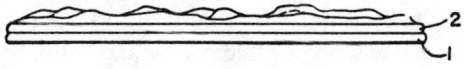

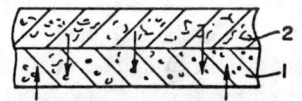

Fig. 6.8 Microwavable pizza product. Cracker dough crust (1) underlies pizza dough (2) to absorb moisture during microwave heating of pizza (Bone and Monoski, 1981).

with ingredients falling within the general limits for pizza dough will remain crisp when microwave baked. It is not at all clear that these methods will provide any browning or will affect flavor. Some of the browning techniques mentioned in the chapter on browning might provide an answer.

**Dinners**

There is little evidence that frozen dinners need to be especially formulated for use in microwave ovens. A dinner designed according to present practice; i.e., with sauces having good freeze-thaw characteristics, and vegetables that retain their texture when frozen and thawed, will perform reasonably well in the microwave oven, provided the precaution is taken of covering the food during heating to prevent excessive dehydration.

The variety of plated dinners being marketed by various food manufacturers is impressive. It is much more impressive when the variety produced in Europe and the United Kingdom are added to the list. There seems to be no limit to the dinner possibilities.

Initially, in the United States, some dinners were plated on filled polyester dishes, lidded with foil and covered with a polyethylene dome. For conventional heating the dome was removed and the foil was left in place. For microwave heating the foil was removed and the dome replaced. The high cost of the filled polyester dish resulted in its eventual replacement with a much lighter polymeric plate and the elimination of the foil cover since the new plate was not dual ovenable. This is one example of the recognition that the microwavable market can stand on its own, and dual ovenability is no longer an important marketing concern. Others are packed in crystallized polyethylene terephthalate (CPET) formed, multicompartment trays with film lids; molded pulp compartmented trays; or polyester coated paperboard trays with lids of the same material. Still other foods are packed in boilable pouches. Lately there has been a shift toward less costly as well as minimal packaging because of the concern that excessive packaging contributes to the waste management problem.

There is, to this writer's knowledge, no literature dealing with the design of dinners for microwave heating in spite of the fact that there are a variety of frozen dinners, casseroles and entrees for microwave and conventional oven preparation on the market. Some general guidelines for reheating chilled foods have been published (Ohlsson and Thorsell, 1984) that are based on experiment and are constructive and useful to review. Other work related mainly to plating arrangements for hospital foodservice appeared in Copson's book (1975). This information is generally applicable if one remembers that the bridge between the frozen state and the thawed condition must be crossed with caution if successful results are to be realized.

Several general guidelines should be followed when designing dinners or casseroles. These were drawn from the work of Ohlsson and Thorsell (1984):

1. Microwave compatible dishes should be used.
2. The shape of the food is important to heating uniformity.

3. Microwave penetration is limited.
4. Heating rate varies with food type.
5. Components of dinners heat at different rates.

**Microwave compatible dishes.** This subject has been discussed previously, but a few additional comments here are in order. If plastic dishes are chosen, they must be able to tolerate the highest temperatures that are apt to occur; for example, the temperature of fats under certain circumstances could easily exceed the distortion temperature of many commonly used plastics. Where dual oven capability is desired, one should not ignore the possibility of using metal containers for all of the reasons discussed before (See Chapter 4).

**Geometry.** Spherically and cylindrically shaped foods tend to concentrate microwave energy in the center when the diameter is in the range of 20 to 60 mm (0.8 to 3.0 in). For larger diameter foods conduction heating will play a role, and time and power must be adjusted to permit completion of conduction heating. Ohlsson and Thorsell (1984) recommended that gravies and sauces be heated separately in paper cups with diameters less than 60 mm, and the cups should not be more than two thirds full. Gravies and sauces can then be poured over the appropriate dinner component.

**Microwave penetration.** The speed of microwave heating is due to its deep penetration into foods. Penetration differs markedly between frozen and non-frozen foods. Where one or two passes of microwaves may occur through non-frozen foods, there are numerous passes through frozen foods before effective heating takes place. Frozen foods are extremely transparent to microwaves until the temperature approaches 0 C, where specific heat and dielectric properties both increase to make food much more lossy. At this point there is a strong tendency for surface heating to dominate and for runaway heating to occur. Recognition of these factors leads to selection of food shapes and thicknesses that tend to give more uniform thawing and heating results. Also since frozen foods are more transparent, a higher power level may be employed initially to raise the temperature more rapidly to just below 0 C; for example, $-2$ to $-4$ C, where specific heat and dielectric loss are just beginning to increase significantly. The heating rate then should be reduced to allow the food to pass through the critical temperature range for complete defrosting before microwave power is increased once again to higher or full power. This capability has been demonstrated in an experimental vending machine (See Chapter 1) and is now available on many consumer ovens.

**Heating rate.** Different foods heat at different rates in microwave ovens. Dense foods are more difficult to heat than less dense foods. If the power level is too high, edges will overheat severely. Although reducing the power level will increase the heating time, the results will generally be more uniform. Most microwave ovens now come equipped with means for programming the microwave power over a wide range of settings as well as programming changes in power. For example, in heating a frozen prepared food item it is possible to use high power to advantage in an initial tempering step then change to a lower power level to finish heating to serving temperature. It is not too far fetched to anticipate that microwave ovens will have the means to read and input coded heating instructions from a bar code on the package. This will place the burden on the product developer to develop accurate programming information to insure customer satisfaction in their microwave heating results.

**Differential heating.** Multi-component dinners present somewhat of a problem in microwave heating, particularly when the components have widely different heating properties. A not uncommon result is for one component to be overheated while others are inadequately heated. One solution, of course, is to select components that heat at the same rate. An alternative is to use devices that control the heating of each component so that a much wider choice of components is possible. Such devices and techniques were discussed in the chapter on packaging. Another alternative involves a unique waveguide feed design by the Berstorff Corporation in which energy is applied at different rates to each component. The procedure requires the precise alignment of the food container on the conveyor belt as it travels between waveguides located above and below the belt. The large compartment is irradiated by one waveguide from which the power is raised to a preset level, held at that level for a period of time, then reduced as the container passes. The other two compartments are irradiated in sequence as the container passes a parallel waveguide with the power level adjusted up or down as the second of the two compartments pass by. Each side of the container may be irradiated by a number of waveguides as it moves down the oven. Thus the total power required is applied in increments until the proper temperatures are reached.

## Casseroles (shelf-stable)

At least one source, cited by Vrabel (1988), predicts that in 10 years the market for shelf-stable entrees will be $1 billion. Vrabel, a spokesperson for Hormel, does not feel that these products will have a serious impact on the frozen food market (Fig. 6.9). Others are not so sure. There are some who feel that shelf-stable foods are not as tasty as frozen foods. Earlier introduction of shelf-stable foods in the retort pouch was not successful, but the microwave oven was not then a factor. In any case, convincing the consuming public to favor shelf-stable over frozen, refrigerated or fresh will depend, as Vrabel explains it, on the "three overriding issues that drive consumer purchase motivation: quality, convenience and value."

The consumer is driven by convenience much more today than ever before because of the microwave oven. Microwave heating of shelf-stable foods is 3 to 4 times faster than frozen foods. Only instant preparation could be more convenient.

# PRODUCT DEVELOPMENT FOR THE CONSUMER MICROWAVE MARKET 147

Fig. 6.9 Shelf-stable microwavable food product.

Value, that is, cost, currently is equal to or less than frozen and might be expected to improve substantially with competition because of the far better logistics of storage, distribution and marketing shelf-stable over frozen or refrigerated foods.

Hormel has its line of Top Shelf shelf-stable entrees made especially for the microwave oven, and was also in the process of introducing their Microwave Ready entrees in Omni containers. Their line of New Traditions breakfast sandwiches has been expanded to 8 items. Others in the shelf-stable market for the microwave oven include Dial Corporation with their Lunch Bucket line, Del Monte's Vegetable Classics and General Foods' Impromptu line, the latter now under Kraft General Foods since Kraft was acquired by the Philip Morris Company.

General Foods had been test marketing the Impromptu line as far back as 1986. These 9 to 11 ounce shelf-stable entrees for the microwave and conventional oven included: Spaghetti and Meat Balls with Tomato Sauce, Swiss Steak and Gravy, Salisbury Steak and Mushroom Gravy, Manicotti in Marinara Sauce, Beef Stroganoff with Noodles, Lasagna with Meat Sauce, Chicken Breast a la King with Rice, Chicken Oriental, Chicken Cacciatore with Noodles and Coq au Vin.

Many other food firms are likely to follow in this market of shelf-stable food products for the microwave oven, at least in the near term. Only time will tell if microwavable shelf-stable food products can compete successfully with their frozen and refrigerated counterparts. All of the shelf-stable products marketed to date have been processed by conventional retorting methods. The processes used vary in the time the product is exposed to heat. For example, a 10 ounce pouch given an $F(0)$ of 6 would be exposed to heat for 31.7 to 47.3 minutes depending on the type of retort, vertical or horizontal, used, whether water or steam/air cooked, and whether agitated or not (Hardt-English, 1982).

The concept of using radio-frequency energy in a process to produce shelf-stable foods was first suggested by Jackson (1947). Although reference was not made to high temperature short time (HTST) processing, the obviously much shorter process time possible with such energy sources would describe at least an HTST-like process. Some references can be cited to actual microwave frequency experimentation. Jeppson (1964) heated skim milk in glass jars to 121 C (250 F) and higher at 915 MHz. Recognizing the possibilities of a microwave assisted sterilization process (Fig. 6.10), patent applications were filed in the same year (Jeppson and Harper, 1967, 1968). Landy (1965) also had patented a process for sterilization in constant volume vapor tight sealed glass tubes and spiral wound 2 mil thick Mylar plastic tubes (Fig. 6.11). There is no evidence that either of these processes were reduced to practice. Landy claimed that steam generated by microwave heating of moisture present or added to the food provided the temperature and pressure conditions for sterilization. Although demonstrated in an 800 watt microwave oven, Landy noted that a conveyorized microwave oven would be needed for commercial production.

Research in sterilization of foods using microwave energy has been carried out and reported on by several organizations, notably The U.S. Army Natick Research

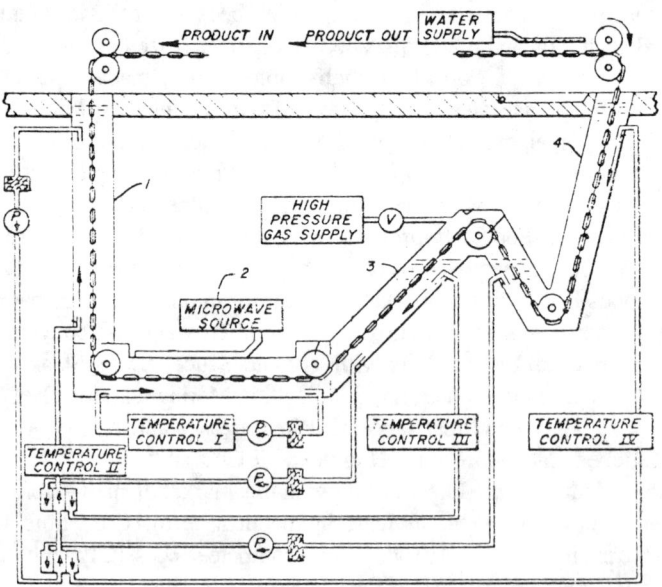

Fig. 6.10 Microwave heating of food products under hydrostatic pressure. Pouches in a continuous link enter through a hydrostatic leg (1) into a section where they are microwave heated in a temperature controlled bath by microwave source (2) before passing through a temperature holding section (3) and a cooling section (4) (Jeppson and Harper, 1967).

# PRODUCT DEVELOPMENT FOR THE CONSUMER MICROWAVE MARKET

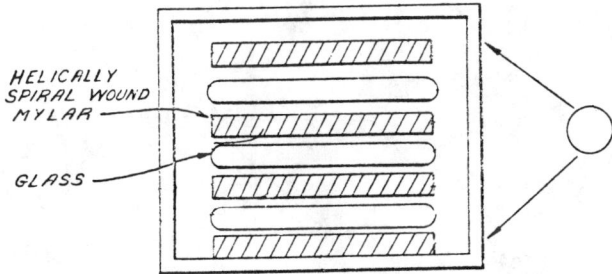

Fig. 6.11 Sterilizing foods in sealed container (Landy, 1965).

& Development Laboratories, Alfa-Laval AB, the Swedish Food Institute, and the U.S. Department of Agriculture. A review of this work is presented in the following paragraphs.

## Research at Natick Laboratories

Kenyon (1970), at the U.S. Army's Natick Laboratories in Natick, Massachusetts, suggested that the shorter process time possible when using microwave energy should result in a product with greatly improved texture and flavor. His interest was in developing improved military rations and led to the design and fabrication of a microwave processing unit to verify this possibility (Fig. 6.12). A fiberglass-reinforced epoxy pipe was installed in a 10 kW, 2450 MHz microwave oven and provided with valve means for introducing pouches into the pipe, conveying means through the pipe and a cooling receiver at the opposite end to receive the processed packages for periodic removal from the pressurized system. Pouches made from a laminate of mylar-polyethylene-polyisobutylene were used in the preliminary evaluation of this concept, fully recognizing that an improved barrier packaging material would be required eventually. Overriding air pressure was provided in the system to balance the internal pressure generated in the package during processing. Temperature measurement of the product during processing was obtained by using paper strip thermometry. The process time varied from 9 to 14 minutes consisting of a heating phase of 4 to 6 minutes, a 3 minute hold and 2 to 5 minutes cooling. The long process was due in part to an initial product temperature into the system of 23.8 C (75 F), rather than a hot fill as in normal processing practice, as well as some time delay due to the length of the end load sections in the unit required to confine microwave energy.

Work to refine the process was continued by Ayoub et al (1974) using Kenyon's equipment. Pouches were wrapped in insulating paper to reduce heat loss to the processing environment. A method of time/temperature integration based on the Maillard reaction was developed to better understand the changes taking place in the system. Small plastic pouches containing 0.5 mL of a tryptone-dextrose solution were placed in the geometric center of the product, and the color developed during processing

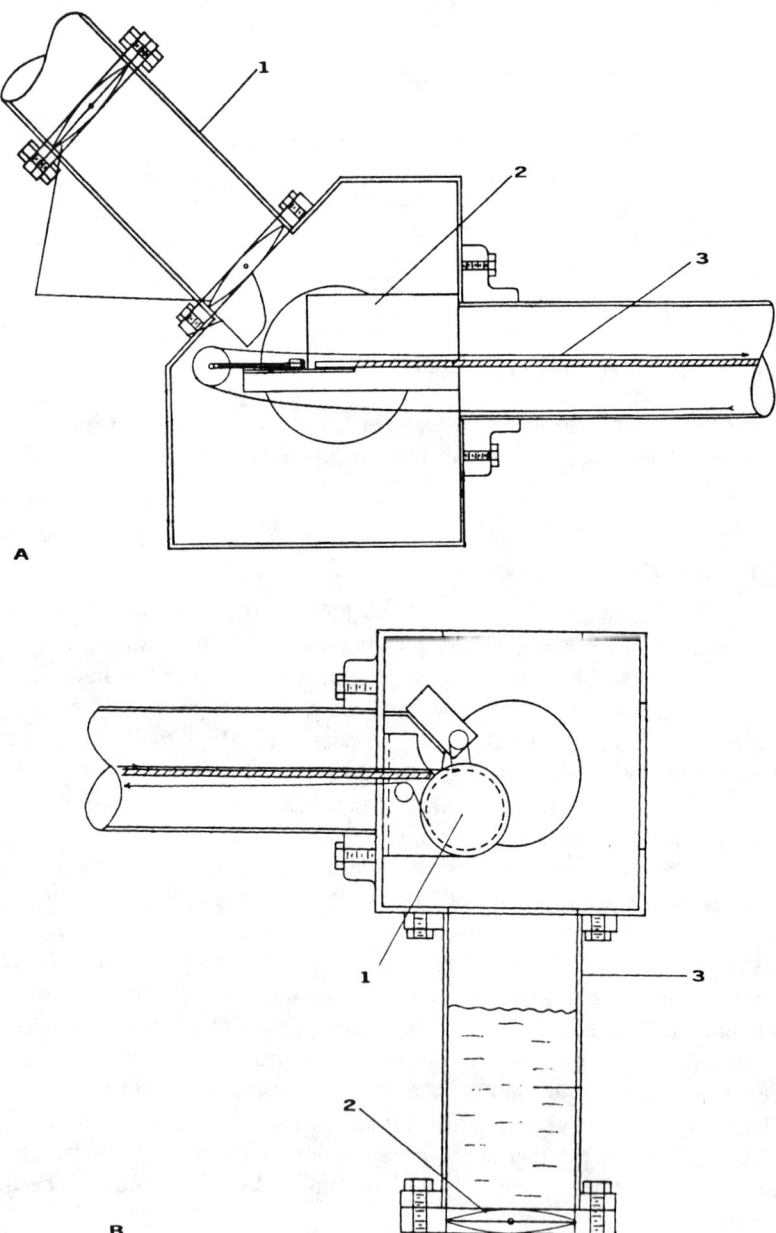

Fig. 6.12 A. Loading end of Natick Laboratory microwave sterilizer. Pouches are loaded through pressurizable Section 1 and after lower butterfly valve is opened, drop onto belt 3 that carries them through the microwave field (not shown). B. Unloading end shows conveyor belt that carries the pouches into the opposite end section where they fall into a cold water tank for cooling and where they are later removed by draining the tank through valve 2 (Kenyon, 1970).

measured with a colorimeter and compared with a standard curve obtained by heating aliquots of this browning solution for increasing periods of time at 121 C. This provided a measure of the F value in the center of the food products processed. Paper thermometers were used to evaluate the temperature distribution throughout the pouches and confirmed that the slowest heating point was in the center as with conventional heating processes.

### Research at Alfa-Laval AB

Somewhat earlier, in 1967, a microwave sterilizing process was under development at Alfa-Laval AB, in Tumba, Sweden. Stenstrom (1982), in commenting on the development of this process, noted that the work had reached an impasse. Differences of as much as 215% between surface and center temperatures of 20 mm thick food slabs were being obtained that required a reexamination of microwave heating technique. They found that if the slab were surrounded with a border of water, microwaves would not see the corners and edges of the food slab as boundaries. The result was a reduction in temperature difference to only 10 per cent.

Stenström (1972) reported on their preliminary results with HTST microwave sterilization of foods and pointed out the problem areas: packaging materials (i.e., adequate barrier properties); cooling (the benefits of rapid microwave processing are lessened if cooling is prolonged) and excessive surface heating. Meat balls, fish fillets, green beans and other unspecified foods were microwave sterilized and stored at room temperature for nearly two years without microbiological alteration. Several patent applications filed in 1971 were granted (Stenström, 1974a,b,c).

In the patents assigned to Alfa-Laval AB, the concept was expressed of using hydrostatic columns in series for preheating and to provide the pressure and temperatures required during the microwave heating step followed by cooling in a second set of hydrostatic columns (Fig. 6.13). Actual food experiments were carried out in a much smaller device called a Microclave in which all of these steps could be performed in series. Some 5000 packages of more than 100 different foods were processed in this unit that demonstrated high quality products could be produced. Several series of inoculated runs were made that proved sterility could be obtained.

In Stenström's patent (Stenström, 1974b) he recognized that a substantial positive temperature gradient from the edges toward the center will occur during the microwave heating step. His patent describes the use of an immersion cooling step just prior to the microwave heating step (Fig. 6.14). With this expedient, the surface will have a greater temperature differential to bridge than the center so that much smaller temperature differences as well as more uniform quality will result.

### Research at the Swedish Food Institute (SIK)

Also during this time period, the early 1970s, a great deal of pertinent research was being carried out at the Swedish Food Institute (SIK), in Goteborg, Sweden,

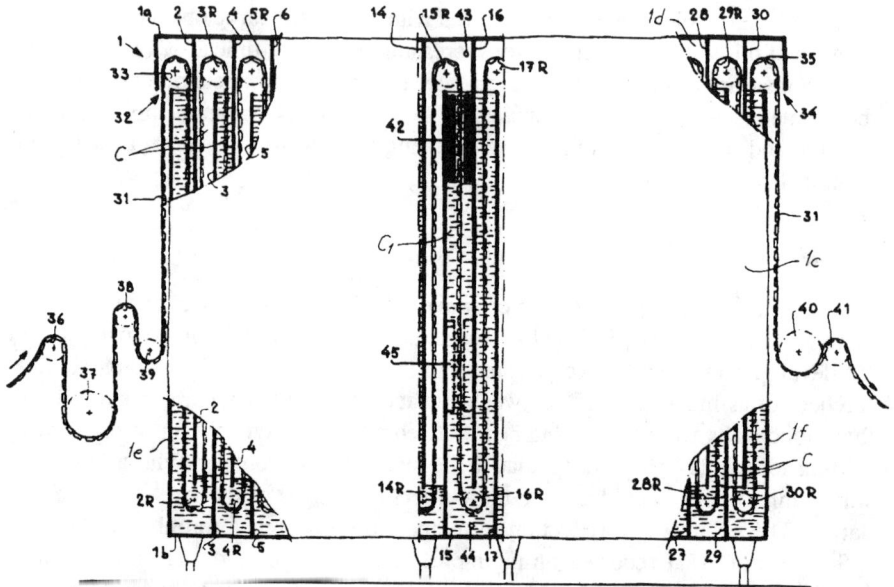

Fig. 6.13 Patent illustration of the Alfa-Laval microwave sterilizer concept (Stenström, 1974b).

research essential to a proper understanding of microwave processing in general. In the first two of a number of papers published by this organization (Risman and Bengtsson, 1971; Bengtsson and Risman, 1971) a simplified method of measuring the important parameter, dielectric loss properties of foods was described and dielectric data on a variety of food products was reported. Their data was compared with data developed much earlier with much more sophisticated apparatus (von Hippel, 1954) and found to be in general agreement. The temperature range of these measurements, however, was only from −20 C to +60 C and only one frequency, 2800 MHz, was used. A later paper by Ohlsson et al (1974) extended the measurements to include 900 and 450 MHz. They concluded that dielectric loss showed an increase with decreasing frequency as well as an increasing temperature dependency. Thus, at least in this range of temperatures, the effect of frequency on the temperature distribution in a material is quite similar; that is, the differences are minor. Ohlsson and Bengtsson (1971) with some of the dielectric data developed at SIK obtained good agreement between computer simulations and actual heating results with beef, ham and simulated meat. The effect of thermal conduction was taken into account in all of these simulations.

In a paper published in 1975, Ohlsson and Bengtsson (1975) reported dielectric measurements of foods in the temperature range of 60 to 140 C. These measurements were necessary in order to be able to draw conclusions on the feasibility of microwave sterilization of food. The same equipment used for earlier measurements was used except that modifications were made to permit operation above the boiling point of water. The results clearly showed that the dielectric loss factor increased with

# PRODUCT DEVELOPMENT FOR THE CONSUMER MICROWAVE MARKET

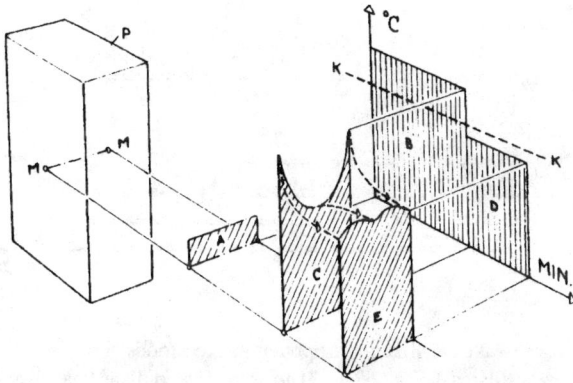

Fig. 6.14 Cooling step prior to microwave heating reduces the temperature gradient from edge to center of the product. Item C shows the high edge temperature of the product P from the preheating step. Item E shows the effect of rapid water cooling on the edge temperature of Item C. When the microwave heating step is completed the temperature across the product at M-M will be more even (Stenström 1974b).

temperature particularly at the lower frequencies and for salty foods. Thus, although an advantage is shown for the lower frequencies in terms of penetration below about 50 C, at temperatures above 100 C, penetration, at least in the case of beef and gravy, is clearly greater for 2450 MHz than for 915 MHz. In the case of distilled water in this range of temperatures 915 MHz penetration is in the range of 700 to 900 mm compared to 2450 MHz where the penetration is below 100 mm. They concluded that 915 MHz would be at a disadvantage in microwave sterilization because of the tendency to enhance surface heating. Except in the Multitherm system where the product is heated while immersed in water some auxiliary heat source still is necessary to prevent heat loss by radiation from the food packages. An alternative to a supplementary heat source was suggested by Kenyon (1976); that is, the use of conveyor belts of insulating material and conveying the microwave heated products between a pair of such belts for a sufficient time to complete the thermal process (Fig. 6.15). The product then must be cooled before it can exit from the pressurized processor. An impingement type air cooling system would appear to be an alternative to water immersion or a cold water spray to bring the product temperature below 100 C. Agitation would be an asset if the product would not be adversely affected. Liquid nitrogen and forced air cooling have been used.

## Research at Western Regional Research Center, USDA

Because of processing restrictions imposed by the microwave system used at the Natick Laboratories, an arrangement was made with the U.S. Department of Agriculture's Western Regional Research Center to design a microwave processing

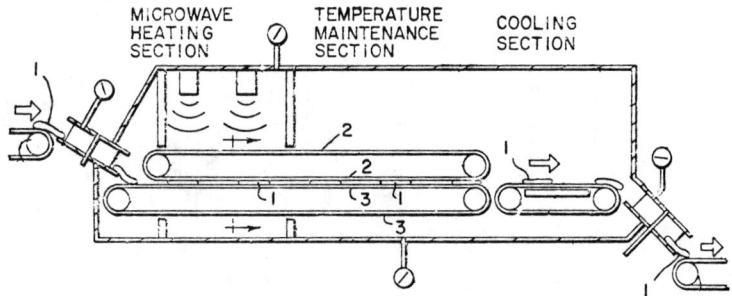

Fig. 6.15 Microwave sterilization of pouch packed foods. Pouches (1) are carried between insulated belts (2 and 3) to minimize radiant heat loss (Kenyon et al, 1976).

system that would eliminate delays between the processing steps. O'Meara et al (1977) constructed a batch type microwave retort (Fig. 6.16) with means for preheating pouches of food with water to near 100 C thus simulating a hot fill; raising the pouches out of the water into a microwave region where the pouches would be heated to processing temperature by rotation in a helical manner past a microwave antenna; followed by reimmersion in water at process temperature to simulate a hold; then cooled by replacing the hot water with cold water to lower the product temperature as quickly as possible to below 100 C.

Using this apparatus O'Meara et al (1977) processed ground beef samples inoculated with P.A. 3679 [$2 \times 10^4$ spores/100 grams]. The one cm thick slabs were vacuum sealed and preheated to 93 C followed by microwave heating at 2 kW for 60 seconds and a 121 C water hold for 2, 6, and 15 minutes. The samples were incubated for 90 days at 37C. Results showed gas formation in all but the 15 minute, 121 C hold samples. Based on these data a number of entree items were processed for 75 seconds at 2 kW followed by a 7 minute hold at 121 C. and compared with a 40 minute 116 C hot water process. From a quality point of view, these microwave processed products approached the quality of frozen entrees. With these results as incentive it was decided to carry out an inoculated pack study to evaluate the efficacy of the process.

Chicken fricassee was selected as the test product for inoculated pack processing. Spores of P.A. 3679 were used as the test organism and inoculated into pieces of chicken meat to give an inoculum of $10^4$ spores. The pouch packaged products were preheated in 95 C water for 20 minutes, then microwave processed for 0, 50, 65 and 75 seconds; held in 121 C water for 0, 5, 10, and 15 minutes; then cooled to below 100 C in 2 minutes. All samples were incubated at 37 C and observed for gas production. With the exception of samples processed for 65 seconds and held 15 minutes at 121 C, all showed gas production.

When the test was repeated using a 20 minute preheat in boiling water followed by microwave heating at 1 kW for 150 seconds — the longer exposure at the lower

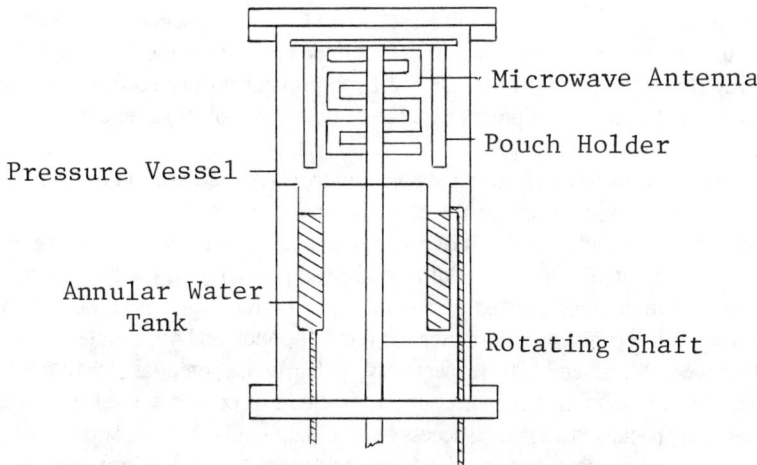

Fig. 6.16 Laboratory microwave sterilizer provides means for preheating, microwave heating and holding product in sequence (O'Meara et al, 1977).

power would allow some equalization of the heating effect — gas formation was still observed in all samples held for up to 12 minutes and some of those held for 15 minutes. Others heated at 2 kW for up to 90 seconds followed by holding at 121 C for up to 15 minutes resulted in one sterile pouch after 90 seconds and a 10 minute hold; no sterile pouches after 75 seconds and a 12 minute hold; three sterile pouches after 75 seconds and a 15 minute hold; two sterile pouches out of five heated for 65 seconds and held 12 minutes; three sterile pouches out of four at 65 seconds and a 15 minute hold.

These researchers concluded from the somewhat erratic results that a serious energy distribution problem existed in their apparatus such that the energy absorption predicted was not being realized. Further energy pattern studies verified this conclusion. That such a poor heating pattern could occur with the elaborate exposure procedure being employed suggests that the solution to this problem may be more complex than anticipated. No inoculated pack data on microwave sterilization of food in microwavable packages has been published since this study was carried out although references have been made to successful sterilization of inoculated packs.

## Some concluding comments on microwave sterilization

Alfastar, a joint venture of Alfa Laval AB and Akerland and Rausing in Sweden was established to bring the high temperature short time microwave assisted process, Multitherm, developed by Alfa Laval (Stenström, 1974a,b,c) to market (Anon, 1983). Although very high quality shelf-stable products have been produced by this process, its technical success depends, obviously, on several factors aside from the fact that the quality must be superior to conventionally retorted products because of the much less severe heat treatment. First, a barrier packaging material that can tolerate processing temperatures at least to 140 C (284 F); second, processing conditions that assure these temperatures are attained throughout the food product mass; third, rapid cooling to below 100 C (212 F) to prevent further cooking; and finally, microwave processing equipment capable of doing the job at acceptable production rates.

Ohlsson (1987) summarized the state of microwave sterilization to date. First he made a comparison of canning, retorting foil pouches and microwave sterilization of food in plastic pouches in terms of process time and C-values (C-value is a measure of sensory and nutritional quality of a food product owing to the heating process). The microwave process showed an advantage in both categories (process time was 3 minutes for microwave, 13 minutes for the foil pouch and 45 minutes for the can; C-values were 28, 65 and 180, respectively). Clearly the potential is with microwave sterilization. Ohlsson also pointed out the problem of corner and edge overheating in rectangular packages and how success in eliminating this problem was demonstrated at SIK when microwave heating food pouches were immersed in water. A typical program of 2.5 minutes of microwave heating at 2450 MHz and a 45 second hold resulted in an F(0) of 6.

A microbiological evaluation was carried out using *Clostridium sporogenes* spores inoculated into peas and chicken a la king and processed at 5 different F-values. On the basis of surviving spores it was shown that the biological F-values exceeded the target F-values indicating bacteriologically safe conditions. It should be noted here that all of this previous work has been with pouches or rigid plastic wrapped slabs of product. There have been no reports in the literature of microwave sterilization of entrees in plastic casserole containers such as are being conventionally retorted and are in test market. One would anticipate that such containers because of their oval shape would minimize the corner heating effect of rectangular slabs, but the gradient from periphery to center would still remain. The significant air space above the product would also impede cooling.

There was some test market activity in Sweden in 1988 with food products pasteurized by the Multitherm process. A commercial installation of this process at BOB Industries, Sweden was reported and another installation was scheduled at Hilldown, Ltd., a leading canner in the United Kingdom (LeMaire, 1985). In 1991, Alfa Laval was acquired by Tetra-Pak and the BOB Industries use of the Multitherm process discontinued. One of the reasons given was the excessive cost of the packaging operation. The products were pasteurized in a rigid, rectangular film wrap then packaged in a coated paperboard tray with a coated paperboard lid.

Batch microwave sterilization equipment suitable for laboratory studies has been offered by Cober Electronics in the United States, OMAC in Italy and Berstorff Maschinenfabrik in West Germany (see Appendix A). Production systems other than the Multitherm process are said to be on the drawing boards and at least one production installation is operational (Fig. 6.17).

There is, therefore, some evidence that the microwave sterilization process can produce a measurably higher quality product than conventional retorting. Whether the improvement is sufficiently better that the consumer will opt for the microwave product remains to be seen. It may be possible to process some products in this manner that defy conventional processing; for example, the processing of a multi-component dinner.

## Some final thoughts on developing microwavable food products

The effect of various physical factors and microwave oven factors on microwave heating performance were covered in Chapter 2. The product developer should be conversant with these factors to the point that they become second nature. With this knowledge as a background product development can be carried out in a more systematic manner. It should prove useful in making over an existing product into a microwavable product as well as for innovative product development; i.e., product invention.

We are conditioned by centuries of tradition to prefer foods prepared in a certain manner. Often these familiar qualities can be preserved by various methods and revived by microwave heating. Indeed, microwave heating has made it possible to produce a very large variety of familiar and special recipe dishes. Witness the growth of ready meals in most industrialized countries in recent years, a growth rate that is directly attributable to the sales of microwave ovens. Many of these ready meals are offered in the frozen state, and there is evidence of substantial growth also in the production of chilled and shelf-stable ready meals. Ready meals is the generic term used, mainly in Europe, to identify frozen, chilled, and shelf-stable prepared dinners or entrees. A great many of these now carry a microwavable symbol on their label, a wavy line beneath the word microwavable, somewhere on the front of the package. This symbol is used in recognition of the importance of identifying a product that is microwavable to the rapidly growing number of microwave oven users. Elsewhere on the label there will be found microwave heating directions.

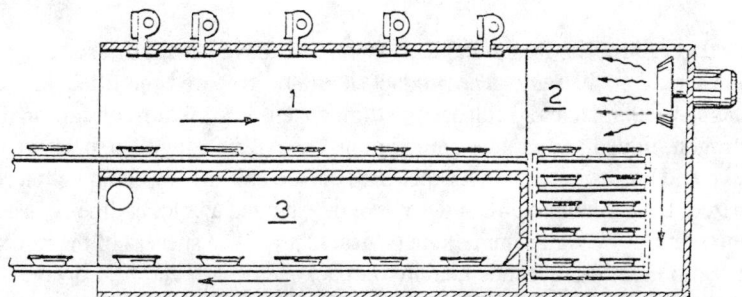

Fig. 6.17 Production microwave sterilizer. Packaged product moves through microwave and thermal heating zone (1) into a temperature holding zone (2), where it is lowered and moves into a cooling zone (3). Entrance and exit air locks to the left are not shown (Guarneri and Ferrari, 1991).

Comments in the following paragraphs are related to developing microwavable food products from the point of view of microwave technology. It is assumed that the developer has identified a product for which there is a potential market and would establish a development effort to produce product for a market test. In some cases the objective will be to produce an existing product in a microwavable form. In other cases the requirement may be to develop a product from scratch; for example, a microwave puffable snack, a cookie, or some other microwave baked item, to name a few. In the first case the developer already has a standard against which to compare the quality of the microwavable product being developed. In the latter case the developer may be breaking new ground. Both of these cases will be addressed.

**The first case: A microwavable version of a ready meal**

The standard against which the microwavable product will be compared, whether frozen, chilled or shelf-stable, probably will be the frozen product. It should be thawed and heated conventionally with care when making the comparison. For some products heating in a pouch in boiling water would be acceptable in which case the product should be vacuum packaged in a flexible pouch to provide the shortest heating time possible. For certain other products, because of the need for browning and crisping, a hot oven is necessary in which case the product could be in a foil or even a tempered glass or ceramic container.

**Frozen version.** Initially, the developmental product should have the identical formulation of the control. It should be packaged and frozen in a configuration that maximizes uniform microwave heating. If a single item, for example, a stew, it should be packaged in a rectangular, square or oval shaped container with generous radii in the case of the first two shapes. The sides of the container should be vertical or nearly vertical. Container material can be chosen from those covered in Chapter 4. Controllable variables include: thickness, weight and to some extent, dielectric properties.

**Chilled version.** The frozen standard identified above should be used and the formulation should be the same. The product should be prepared and filled into suitable containers, vacuumized and released with nitrogen or a mixture of carbon dioxide and nitrogen, lidded and rapidly chilled to just above freezing. The packaged product also could be pasteurized in a water bath before chilling; or could be microwave pasteurized, by a combination of microwave heating and a holding time at pasteurization temperature (by water immersion or heated air). The success of the microwave pasteurization method depends heavily on package design and microwave heating rate, a method that still is under investigation. Obviously, a package design with shielding means that prevents excessive peripheral heating would be desirable. This can be made a part of the package or built into a holder that carries the product through the microwave field.

**Shelf-stable version.** The control, packaging and product formulation should be the same as above. Sterilization can be accomplished in a variety of pressurized systems currently being used for such products on the market. This is the most severe treatment of a product, since the product is exposed to temperatures of 250 F or higher to insure that all harmful microorganisms have been killed. A process based on microwave sterilization has demonstrated that product quality can be significantly improved over conventional sterilization methods. Some microwave sterilized products have appeared recently on the European market. It is too early to determine if they will establish a niche in the marketplace.

Obviously, to develop a microwave sterilized product requires a pressurized microwave system. At this time such a device is not a catalog item. A few have been built over the years for research work and demonstrated that superior quality products could be produced. It is possible to have a unit custom built and there are several firms in the USA, the UK, Europe, Australia, Japan and South Africa with the necessary skills to build such a unit. A list of microwave equipment manufacturers can be found in Appendix A.

After development samples have been prepared by one or all of the above methods, the next step is to compare the reheated product against the control. This introduces an uncontrollable variable, the microwave oven. No microwave oven has a perfectly uniform microwave energy distribution pattern and no two oven brands are identical. They differ in heating pattern and power level, which is why most microwave heating directions give a range of heating times.

Product developers should never forget that control over a product's performance is completely lost when a product is sold to a consumer. This applies particularly to frozen products that may undergo changes in temperature and quality between the checkout counter and the home freezer. An effort should be made in product development to compensate, if at all possible, for this injury. Shelf-stable products are less prone to such insults. Chilled items have a much shorter shelf life and should be used immediately for best results. Marks & Spencer in the United Kingdom, for example, have some chilled items with a very short "sell by" date; as short as 4 days for certain poultry products. These products, however, have not been pasteurized; indeed some of them probably would defy any such thermal treatment and many may not have been identified as microwavable. The quality depends on very high microbial standards and a dedicated delivery system that tolerates no breach of the established temperature standards.

Quality Control should insure that all ingredients used in the formulation meet tight specifications for bacterial contamination and that samples are checked at all points in the process where recontamination could occur. Finally, a percentage of the products should be incubated for a period of time to verify that the product meets the microbiological specification that has been established to allow it to be released to the marketplace.

## The second case: Innovative product development

It is not essential that all microwavable products be in a heatable package. A product could be similar to a baked potato; that is, one or two could be prepared at a time. It could be a cookie product such as that described by Boehm and Fazzolare (1990), a filled cookie that when microwave heated the filling expands and exudes out of the top of the cookie and flows down the outside like a volcano. The development required a formulated shelf-stable filling that would be a good microwave energy absorber and that on heating would change to a flowable mass. As described in the above patent, a two phase filling was developed: an oil based filling and a water based filling. The cookie is produced by co-extruding a dough and the oil based filling. The dough piece is then baked whereby there is formed a hollow interior section. The hollow interior is filled with the water based filling. When heated in a microwave oven the water based filling expands and exudes out through the top of the cookie, flowing down the exterior and coating it.

The example of pancakes and syrup described earlier is another example of innovative development. Several other examples were cited earlier in this chapter. Patents, however, do not always describe successful developments. If they did there would be a number of successful microwavable french fried potatoes on the market to challenge the Fast Food industry product. A sampling of microwavable french fried potatoes on the market provides ample evidence that there is still much room for improvement.

To be innovative in product development requires an objective. A new snack item would be a reasonable objective for discussion purposes. To be different let this snack item be vegetable rather than cereal or potato based. Further let it be a puffable vegetable served warm with a variety of seasoned coatings. Let it be a carrot based snack.

It is known that diced carrots when air dried will at some moisture level become case hardened; that is, a hard skin will form as the moisture movement to the surface diminishes to the point that the surface temperature increases and a skin forms. It is highly likely that this will produce a condition similar to popping corn and at some appropriate moisture level the case hardened carrot dice will expand when exposed to microwave heating. Thus, the first problem will be to determine the optimum moisture level for puffing. Various drying techniques must be explored and samples produced with varying moisture levels for test purposes. Another variable will be dice size. What is the effect of dice size on puffing volume? Since dice will expand to form spherical or near spherical puffed product some thought needs to be given to the package design.

The package must protect the product and could serve as the puffing container as in the case of microwave popcorn. It could also be the holder for shaking the product and coating together. Therefore, some thought needs to be devoted to the volume of puffed carrot dice desired per package.

Storage studies need to be carried out to determine if undesirable changes occur at the moisture level that gives acceptable puffing. Is mold growth likely to occur?

If so what treatment can be applied to inhibit mold growth? Can the coating be applied before puffing; that is, can the coated product be packaged so that application of the coating after puffing is not necessary?

Is a pretreatment of the carrot dice necessary to create a more lossy case hardened product on drying? A soak in flavored, salted water might provide a desirable taste sensation and also affect microwave absorption. What kind of coating would give some crunchiness to the snack? Could a candy-like coating be applied? This would have to be a formulation at least partially transparent to microwaves for microwave heating of the internal moisture to create puffing. The coating must be able to expand with the expansion of the dice.

After a prototype has been developed, taste panels will be necessary to determine product acceptability.

This example has recognized the importance of moisture content, the dielectric loss of the product, dimensions, geometry, weight, packaging. Other examples will suggest themselves to product developers as they become conversant with microwave technology and the microwave literature. A careful, diligent review of the literature, particularly the earlier literature as a background to product development is strongly recommended.

## REFERENCES

Andreas, D.W. and Cox, D.H. (1990). Microwave receptive heating sheets and packages containing them. U.S. Patent 4,943,439.

Anonymous (1982). Microwave oven cake mix introduced by Proctor & Gamble. Food Eng. 54 (12) 30.

Anonymous (1983). Multitherm unveiled. Food Manuf. 58 (12) 23-24.

Anonymous (1985). Research to explore untapped markets for warm snacks. DuPont Pack. 3, 4.

Ayoub, J.A., Berkowitz, D., Kenyon, E.M. and Westcott, D.E. (1974). Continuous microwave sterilization of meat in flexible pouches. J. Food Sci. 39:309-313.

Bengtsson, N.E. and Risman, P.O. (1971). Dielectric properties of foods at 3 GHz as determined by a cavity perturbation technique. II. Measurements on food materials. J. Microwave Power 6 (2) 107-123.

Boehm, D.G. and Fazzolare, R.D. (1990). Filled cookie. U.S. Patent 4,948,602.

Bone, D.P. and Monoski, P.M. (1981). Frozen pizza crust and pizza suitable for microwave cooking. U.S. Patent 4,283,424.

Copson, D.A. (1957). Microwave baking-angel food cakes successfully baked in latest test. Western Baker, September, 22, 25.

Copson, D.A. (1975). Microwave Heating, 2nd Ed. Van Nostrand Reinhold/AVI, New York.

Decareau, R.V. (1975). Developing food products for the microwave oven market. Microwave Energy Appl. Newsl. 8 (1) 3-5, 14.

Fergusson, J.L. (1976). Exciting products for microwave ovens. Food Eng. 48 (9).

Gorfein, H., Rahman, A.R., Cohen, S.H. and Westcott, D.E. (1978). Frozen french fries heated in microwave ovens. Microwave Energy Appl. Newsl. 11 (5) 7-8, 10, 12, 14.

Guarneri, R. and Ferrari, C. (1991). Method for heat prepackaged food products using microwaves in a heated superatmospheric chamber. U.S. Patent 4,999,471.

Haagensen, D.B. (1958). Electronic heating apparatus. U.S. Patent 2,827,537.

Hardt-English, P.K. (1982). Institutional pouches-costs and directions. Activities Rept. *35* (1) 16–23.

Jackson, J.M. (1947). Electronic sterilization of canned foods. Food Eng. *19* (5) 124–126, 224.

Jeppson, M.R. (1964). Techniques of continuous microwave processing. Cornell Hotel & Rest. Admin. Q. *5* (1) 60–64.

Jeppson, M.R. and Harper, J.C. (1967). Microwave heating of substances under hydrostatic pressure. U.S. Patent 3,335,253.

Jeppson, M.R. and Harper, J.C. (1968). Microwave heating substances under hydrostatic pressure. U.S. Patent 3,398,251.

Kenyon, E.M. (1970). The feasibility of continuous heat sterilization of food products using microwave power. Tech. Rept. 71-8-FL, U.S. Army Natick Laboratories, Natick, MA 01760.

Kenyon, E.M., Berkowitz, D. and Ayoub, J.A. (1976). Apparatus for continuous microwave sterilization of food in pouches. U.S. Patent 3,961,569.

Koshida, D., Sigisawa, K., Majimo, J. and Hattori, R. (1982). Method for producing dry fruit chip. U.S. Patent 4,341,803.

Labell, F. and Rice, J. (1985). The retortables-plastics that can take the heat. Food Proc. *46* (3) 50–52, 54–56.

Landy, J.J. (1965). Method of sterilizing food in sealed containers. U.S. Patent 3,215,539.

LeMaire, W.H. (1985). This processing/packaging system uses...microwaves. Food Eng. *57* (6) 50–51.

Lopez-Gavito, L. and Pigott, G.M. (1983). Effects of microwave cooking on textural characteristics of battered and breaded fish. J. Microwave Power *18* (4) 345–353.

Martin, D.J. and Tsen, C.C. (1981). Baking high-ratio white layer cakes with microwave energy. J. Food Sci. *46*:1507–1513.

Mickle, J.B., Smith, W.J. and Dicken, L.M. (1980). Method of manufacturing a high protein snack food. U.S. Patent 4,183,966.

Norris, H.R., Niemand, C.M. and Andreas, D.W. (1976). Precooked farinaceous foods adopted for microwave heating and a syrup topping therefore. U.S. Patent 3,983,256.

Ohlsson, T. (1987). Sterilization of foods by microwaves. Presented at Int'l Sem. on New Trends in Asceptic Processing and Packaging of Foodstuffs, Munich, October 22–23.

Ohlsson, T. and Bengtsson, N.E. (1971). Microwave heating profiles in food. Microwave Energy Appl. Newsl. *4* (6) 3–8.

Ohlsson, T. and Bengtsson, N.E. (1975). Dielectric food data for microwave sterilization processing. J. Microwave Power *10* (1) 93–108.

Ohlsson, T., Bengtsson, N.E. and Risman, P.O. (1974). The frequency and temperature dependence of dielectric food data as determined by a cavity perturbation technique. J. Microwave Power *9*:129–145.

Ohlsson, T. and Thorsell, U. (1984). Problems in microwave reheating of chilled foods. J. Food Serv. Syst. *3*:9–16.

O'Meara, J.P., Farkas, D.F. and Wadsworth, C.K. (1977). Flexible pouch sterilization using combined microwave-hot water hold simulator. Contract No. (PN)DRXNM 77-120. U.S. Army Natick Research & Development Laboratories, Natick, Massachusetts (unpublished report).

Ottenberg, R. (1985). Yeast-raisable wheat-based products that exhibit reduced deterioration in palatability upon exposure to microwave energy. U.S. Patent 4,560,559.

Pinegar, R.K. (1986). Process for preparing parfried and frozen potato products. U.S. Patent 4,590,080.

Pothier, R.G. and Ford, T.E. (1975). Partially shielded food package for dielectric heating. U.S. Patent 3,865,301.

Risman, P.O. and Bengtsson, N.E. (1971). Dielectric properties of food at 3 GHz as determined by a cavity perturbation technique. I. Measuring technique. J. Microwave Power 6 (2) 101–106.

Saunders, F.R. and McLaughlin, R.L. (1980). Potato segment and process for preparing frozen french fried potatoes suitable for microwave reheating. U.S. Patent 4,219,575.

Stangroom, M. (1976). Method of baking a pizza using microwave energy. U.S. Patent 3,975,552.

Stenström, L.A. (1972). Taming microwaves for solid food stabilization. Pres. at 7th Microwave Power Symposium, May 24–26, Ottawa, Canada.

Stenström, L.A. (1974a). Method and apparatus for treating heat sensitive products. U.S. Patent 3,809,844.

Stenström, L.A. (1974b). Heating of products in electro-magnetic field. U.S. Patent 3,809,845.

Stenström, L.A. (1974c). Heat treatment of heat-sensitive products. U.S. Patent 3,814,889.

Stenström, L.A. (1982). Achilles and the Tortoise and Multitherm: two new food stabilization systems. Activities Rept. of the R & D Assoc. *34* (1) 55–67.

Van Hulle, G.J., Anker, C.A. and Franssell, D.E. (1983). Method for preparing sugar coated puffed snacks upon microwave heating. U.S. Patent 4,409,250.

von Hippel, A.R. (1954). Dielectric Materials and Applications. MIT Press, Cambridge, Massachusetts.

Vrabel, J. (1988). Are shelf stable microwave entrees just a passing fad or are they truly the beginning of a new category. Proc. Pack Alim. '88, March 22–24, San Francisco.

# CHAPTER SEVEN

# NUTRITION

## INTRODUCTION

The emphasis today is on better nutrition and nutrition-related labelling on food packages is evidence of this trend. Americans are more health conscious than ever. Products developed for the microwave oven have an advantage over other methods of food preparation for serving in that it can be advertized that "What the label says is what you get." Microwave preparation can be very precise and the nutrient content of the food eaten can be as close as is physically possible to that stated on the label. To quote Alexander Williams, Executive Vice President of the Campbell Soup Company, on nutrition related advertizing and labelling, "The better we can be at communicating the nutritional benefits of our products to consumers — with clarity and credibility — the more effective this tool will be in helping us sell." (Semling, 1984).

Almost any method of processing raw food will have an adverse effect on certain of its nutrients, notably the labile vitamins. However, if the best practices are used in the initial processing, any further degradation will have to come from mishandling in storage and distribution and from the method of heating for service. Except for in-house production there is little control over storage and distribution, but there is control over the method of heating for service.

The immediacy provided by microwave cooking and heating; that is, that foods can be prepared for service much closer to the time of consumption than heretofore possible, could have an important influence on the well-being of a large part of our population. This includes foodservice both at home and away from home. It applies to products developed for the entire foodservice market and could have important implications on the design of foodservice operations.

## REVIEW OF THE LITERATURE

In general, when we speak of nutrition, we think of vitamins. Nutrition, however, includes, in addition to vitamins, minerals and calories from carbohydrates, fats and proteins. Most of our nutrition comes in the form of food and so we must also consider the moisture content of the foods we consume. This review, will be limited to the effect of microwaves on vitamins and moisture content. Excellent, more comprehensive, reviews have been prepared by Cross and Fung (1982) and Lorenz (1976).

## Vitamins

The effect of microwave energy on vitamins in food was studied shortly after the first microwave ovens became available. The work of Proctor and Goldblith (1948) indicated that in most cases microwave energy was less harmful to thiamine and riboflavin in certain meats and fish than other cooking methods (Table 7.1). The greater moisture retention with the microwave method appeared to be an important factor.

Certainly, the higher surface temperatures attained in frying and grilling were responsible for the greater losses reported by these methods. Results in microwave baking of cake mixes showed better retention of thiamine and, with one exception, riboflavin (Table 7.2). The much longer baking times at higher temperatures would appear to be responsible for the higher losses with conventional baking. Longer microwave baking times also affects retention of these vitamins (Table 7.3).

Microwave blanching was shown to have very little effect on the ascorbic acid content of a variety of vegetables (Table 7.4). The blanching studies were carried out at 3000 MHz and at a power level of 2000 watts. Samples weighing 100 grams and in thermoplastic bags were heated for varying periods of time to determine the exposure needed to inactivate the enzymes catalase and peroxidase. The exposure times ranged from 20 to 30 seconds under these conditions. The plastic bags were immediately cooled by immersion ice water. By comparison, blanching in boiling water ranged from one to three minutes and steam blanching two to four minutes. At that early stage in the development of our understanding of microwave heating technology, Proctor and Goldblith suggested that microwave (radar) blanching might have practical advantages in food processing if continuous microwave equipment became a reality.

TABLE 7.1

THE RETENTION OF THIAMINE AND RIBOFLAVIN IN HAMBURG, HADDOCK AND FRANKFURTERS COOKED BY MICROWAVE AND CONVENTIONAL MEANS

| Food | Cooking Method | Moisture Loss (%) | Thiamine (%) | Riboflavin (%) |
|---|---|---|---|---|
| Hamburg | Microwave | 5.0 | 100.0 | 100.0 |
| | Fried | 19.8 | 96.0 | 88.0 |
| Haddock | Microwave | 6.4 | 53.6 | 87.2 |
| | Baked | 19.2 | 67.4 | 92.2 |
| Frankfurts | Microwave | 2.9 | 71.8 | 100.0 |
| | Gibbs oven | 2.9 | 57.6 | 96.0 |

From Proctor and Goldblith (1948)

## TABLE 7.2
## THE RETENTION OF THIAMINE AND RIBOFLAVIN IN CAKE MIXES BAKED IN A MICROWAVE OVEN AND IN A GAS OVEN

| Product | Method | Time (min) | Moisture Loss (%) | Thiamine (%) | Riboflavin (%) |
|---|---|---|---|---|---|
| Devil's Food A | Microwave | 1-5/6 | 7.1 | 90.0 | |
| | Gas | 25 | 13.1 | 77.0 | |
| Devil's Food B | Microwave | 1-5/6 | 6.7 | 78.6 | 96.5 |
| | Gas | 25 | 14.4 | 53.1 | 100.0 |
| Gingerbread | Microwave | 1-5/6 | 8.0 | 92.0 | 68.0 |
| | Gas | 25 | 10.8 | 88.6 | 100.0 |
| White cupcakes | Microwave | 1-1/4 | 8.0 | 100.0 | 75.0 |
| | Gas | 20 | 16.0 | 81.1 | 82.0 |

From Proctor and Goldblith (1948)

## TABLE 7.3
## THE EFFECT OF MICROWAVE BAKING TIME ON THE THIAMINE CONTENT OF GINGERBREAD CUPCAKES COMPARED TO HOT OVEN BAKING

| Time (min) | Oven Type | Moisture (%) | Thiamine (%) |
|---|---|---|---|
| 1.5 | Microwave | 19.0 | 77.9 |
| 1.25 | Microwave | 15.2 | 81.1 |
| 1.0 | Microwave | 10.8 | 87.9 |
| 25 | Gas Heated | 14.8 | 52.3 |

From Proctor and Goldblith (1948)

Some studies with corn-on-the-cob heated in thermoplastic bags were carried out by these researchers. Two ears of corn per bag, heated for one minute at 2000 watts, the bag sealed and cooled by immersion in ice water gave complete blanching and the product was described as having excellent texture and flavor. Similar samples were frozen and found to be acceptable in later taste tests. The plastic package also was found to be presentable as it shrank tightly around the corn. The fact that the process has not been adopted to date can be attributed, at least in part, to the seasonal nature of the process and therefore the need to find some supplementary use of the equipment during the off season to justify the investment in capitol equipment.

Thomas et al (1949) reported on a comparison of various home cooking methods on the retention of several water soluble vitamins in vegetables (Table 7.5 and 7.6). The work was carried out in equipment similar to that used by Proctor and Goldblith.

Their results clearly indicated that considerable amounts of ascorbic acid, thiamine and riboflavin are leached out of vegetables into the cooking liquid and lost as the

## TABLE 7.4
### EFFECT OF BLANCHING ON THE ASCORBIC ACID CONTENT OF VEGETABLES

| Method | Ascorbic Acid (%) | | | | |
|---|---|---|---|---|---|
| | Carrots | Spinach | Peas | Green Beans | Broccoli |
| Microwave | 100.0 | 96.0 | 100.0 | 100.0 | 100.0 |
| Boiling water | 36.7 | 23.5 | 92.8 | 80.0 | 87.0 |
| Steam | 82.4 | 37.6 | 100.0 | 98.0 | 93.5 |

From Proctor and Goldblith (1948)

## TABLE 7.5
### EFFECT OF COOKING METHOD ON CAROTENE AND ASCORBIC ACID IN VEGETABLES

| Method | Time (min) | Carotene (%) In Solids | Ascorbic Acid (%) | |
|---|---|---|---|---|
| | | | Solids | Liquid |
| Broccoli | | | | |
| Microwave | 5 1/4 | 125 | 64 | 23 |
| Boiling | 7 | 130 | 60 | 25 |
| Pressure | 5/6 | 129 | 72 | 6 |
| Cabbage | | | | |
| Microwave | 6 | | 59 | 31 |
| Boiling | | | 62 | 37 |
| Pressure | | | 71 | 10 |
| Carrots | | | | |
| Microwave | 2 1/4 | 118 | 83 | 15 |
| Boiling | 9 | 121 | 80 | 10 |
| Pressure | 1 | 120 | 77 | 13 |
| Potatoes | | | | |
| Boiled | | | | |
| Microwave | 6 | | 81 | 13 |
| Saucepan | 18 | | 76 | 18 |
| Pressure | 9 | | 86 | 5 |
| Baked | | | | |
| Microwave | 2 | | 50 | |
| Oven | | | 47 | |

Adapted from Thomas et al (1949)

cooking liquid is frequently discarded. Although the cooking time for vegetables was not unusually long in the microwave oven, the time did reflect the volumes of water added in the cooking operation. The samples sizes were 352 grams of cabbage, 255 grams of carrots and 330 grams of broccoli to which were added 690, 110 and 450 grams of water, respectively. In effect this was not microwave cooking

## TABLE 7.6
## EFFECT OF COOKING METHOD ON RIBOFLAVIN AND THIAMINE IN VEGETABLES

| Method | Time (min) | Riboflavin (%) In Solids | Liquids | Thiamine (%) In Solids | Liquids |
|---|---|---|---|---|---|
| Broccoli | | | | | |
| Microwave | 5 1/2 | 71 | 31 | 76 | 31 |
| Boiling | 7 | 69 | 33 | 75 | 33 |
| Pressure | 5/6 | 94 | 8 | 90 | 8 |
| Cabbage | | | | | |
| Microwave | 6 | 69 | 19 | 62 | 42 |
| Boiling | 15 | 61 | 35 | 53 | 52 |
| Pressure | 1 1/4 | 95 | 2 | 88 | 3 |
| Carrots | | | | | |
| Microwave | 2 1/4 | 93 | 11 | 91 | 14 |
| Boiling | 9 | 90 | 12 | 88 | 12 |
| Pressure | 1 | 93 | 14 | 85 | 15 |
| Potatoes | | | | | |
| Boiled | | | | | |
| Microwave | 6 | | | 91 | 10 |
| Pan | 18 | | | 83 | 14 |
| Pressure | 9 | | | 92 | 3 |
| Baked | | | | | |
| Microwave | 2 | | | 86 | |
| Oven | 39 | | | 85 | |

Adapted from Thomas et al (1949)

but rather microwave boiling, and in this context it can be more readily understood why the vitamin content of the solids was similar to that for conventional boiling. A number of studies have shown that adding water is not necessary when cooking with microwave energy. The naturally present moisture in foods is sufficient in many cases. This work was confirmed by the studies of Gordon and Noble (1959) who compared boiling, pressure saucepan and waterless microwave cooking of cabbage, broccoli and cauliflower (Table 7.7).

Chapman et al (1960) evaluated microwave cooking of fresh and frozen broccoli in a consumer type microwave oven. One pound quantities of fresh broccoli and two 10-ounce packages of frozen broccoli were cooked by both boiling and by microwave. The broccoli was cooked in covered three quart ovenproof glass casseroles. Ascorbic acid retention varied from 97 per cent for three minutes cooking to 80 per cent for 12 minutes microwave cooking of fresh broccoli; and 98 per cent for 6 minutes to 86 per cent for 15 minutes cooking of frozen broccoli. There were no measurable losses for carotene owing to microwave cooking. Flavor was judged to be about the same for either method and color slightly better for the microwave method than boiling.

TABLE 7.7
COMPARISON OF WATERLESS MICROWAVE COOKING OF VEGETABLES
WITH TRADITIONAL METHODS

| Cooking method | Ascorbic Acid Retention (%) | |
| --- | --- | --- |
| | In Solids | In Liquids |
| Cabbage | | |
| Boiling water | 38 | 37 |
| Pressure saucepan | 70 | 6 |
| Microwave | 80 | 4 |
| Cauliflower | | |
| Boiling water | 73 | 16 |
| Pressure saucepan | 82 | 5 |
| Microwave | 90 | 4 |
| Broccoli | | |
| Boiling water | 45 | 55 |
| Pressure saucepan | 81 | 4 |
| Microwave | 87 | 7 |

Adapted from Gordon and Noble (1959)

Campbell et al (1958) were able to demonstrate better retention of ascorbic acid in vegetables cooked in a microwave oven over pressure or saucepan cooking (Table 7.8). The destructive effect of steam table holding on the ascorbic acid content of broccoli and green peas was evident. The effect of cooking time also was demonstrated (Table 7.9) and illustrated the need for precise control of microwave cooking time to insure optimal retention of water soluble vitamins.

It should be emphasized that this early work was carried out in microwave ovens designed to meet the needs of commercial foodservice operations. The power output of these early ovens was 1.6 to 2.0 kilowatts; thus the heating rates were substantially higher than in today's 600 to 700 watt consumer ovens. Indeed, it is somewhat surprising that the microwave cooking results were so favorable. It is also clear that waterless cooking should result in a greater intake of the naturally occurring water soluble vitamins than when the vegetables are cooked in water by microwave or conventional methods. In addition, the flavor obtainable with microwave cooking can be characterized as fresher. Fresh corn-on-the-cob, for example, is a unique taste experience when cooked the waterless microwave way.

Kylen et al (1961) compared cooking of fresh and frozen vegetables by conventional and microwave means and found no significant differences in the ascorbic acid content. In all but one case the vegetables were cooked in water by both methods so that in effect the microwave oven was essentially cooking the vegetables by conduction of heat from the boiling water. The quantities of water added to the vegetables contributed significantly to the cooking time so that both methods also were similar in the length of cooking time.

## TABLE 7.8
### EFFECT OF COOKING METHOD ON RETENTION OF ASCORBIC ACID IN VEGETABLES

| Vegetable | Method | Time (min) | Retention (%) |
|---|---|---|---|
| Fresh cabbage | Microwave | 5 | 92.75 |
| Fresh cabbage | Pressure | 3 | 50.3 |
| Fresh broccoli | Microwave | 3 | 72.1 |
| Fresh broccoli | Range top | 10 | 59.9 |
| Frozen broccoli spears | Microwave | 3 | 103.2 |
| Frozen broccoli spears | Range top | 13 | 83.1 |
| Frozen broccoli chopped | Microwave | 3 | 97.9 |
| Frozen broccoli chopped | Range top | 11 | 69.5 |
| Frozen peas | Microwave | 1 | 100.0 |
| Frozen peas | Range top | 6 | 70.3 |

Adapted from Campbell et al (1958)

Eheart et al (1964) found no significant difference in the ascorbic acid retention in peas, spinach, broccoli and potatoes cooked with or without boiling water by microwave or conventional means. Details lacking in most of these research studies make it difficult to render meaningful comparisons. It is likely that in most cases the flavor of microwave cooked vegetables were judged against that which was conventionally cooked rather than by some other standard.

Dietrick et al (1970) compared continuous microwave, conventional and combination blanching of Brussels sprouts and found that the combination of microwaves and steam or water effectively inactivated the peroxidase enzyme. They claimed that the process stabilized the flavor and was as effective as conventional blanching in retention of chlorophyll and ascorbic acid.

Lanier and Sistrunk (1979) found no significant difference in pantothenic acid, niacin or total carotenoids in sweet potatoes cooked by boiling, baking, steaming, microwave or canning. There were slight differences in vitamin C and significant differences in riboflavin. In this and the three previous references, details of microwave power level were not provided, or the amount of water used in cooking. It can be assumed, however, that microwave procedures probably were not optimized

## TABLE 7.9
### EFFECT OF MICROWAVE COOKING TIME ON THE ASCORBIC ACID CONTENT OF FROZEN PEAS

| | Time (min) | Retention (%) |
|---|---|---|
| Peas (water added) | 3 | 66.7 |
| Peas (water added) | 1 | 84.4 |
| Peas (without water) | 1 | 95.0 |

Adapted from Campbell et al (1958)

and that could account for results where the retention of vitamins were low for the microwave method.

Armbruster (1978) presented data on the ascorbic acid of fruits and vegetables cooked by microwave and conventional methods and found that in 75% of the cases the ascorbic acid content of microwave cooked products was significantly higher.

Thomas et al (1949) generally confirmed the results of Proctor and Goldblith (1948) with regard to the effect of meat cooking methods on riboflavin and thiamine except that losses were greater with microwave roasting of beef (Table 7.10). They concluded that this was owing to the necessity of overcooking the outside of the roast rather severely to obtain the proper internal temperature. Obviously, conventional roasting techniques are not completely transferable to microwave roasting. Since that time it has been determined that beef can be microwave roasted with results that compare favorably with the best conventional roasting methods. A much lower reference temperature must be used, 90-120 F, compared to 140 F for a rare roast cooked conventionally. In microwave roasting considerable heat is built up in the outer one inch or so of the roast that is then conducted inward after the roast has been removed from the oven (see Chapter 2). Consequently, microwave oven time plus standing time constitutes the total cooking time.

Campbell and Gretchell (1957) in using this technique obtained much more favorable results in the retention of thiamine and riboflavin (Table 7.11). In an earlier study Campbell (1954) reported similar results for pork loin roasts (Table 7.12).

Lushbough et al (1962) in a study of the effect of heating method on meats analyzed beef roasts for the thiamine content in the inner and outer layers. They recognized

TABLE 7.10
EFFECT OF COOKING METHOD ON VITAMIN RETENTION IN MEATS

| Method | Niacin | Retention (%) Riboflavin | Thiamine |
|---|---|---|---|
| Beef Patties | | | |
| Microwave | | | |
| Single | 93 | 103 | 89 |
| Multiple | 89 | 99 | 77 |
| Grilled | 91 | 105 | 55 |
| Pork Patties | | | |
| Microwave | | | |
| Single | 90 | 99 | 91 |
| Multiple | 81 | 87 | 91 |
| Grilled | 84 | 102 | 79 |
| Beef Roasts | | | |
| Microwave | 73 | 84 | 63 |
| Electric oven | 81 | 90 | 75 |

Adapted from Thomas et al (1949)

TABLE 7.11
EFFECT OF COOKING METHOD ON THE RETENTION OF VITAMINS IN BEEF ROASTS

| Method | Thiamine (%) | Riboflavin (%) |
|---|---|---|
| Microwave | 82.6 | 104.8 |
| Electric Oven | 38.7 | 110.7 |

Adapted from Campbell and Gretchell (1957)

TABLE 7.12
EFFECT OF COOKING METHOD ON THE RETENTION OF VITAMINS IN PORK ROASTS

| Method | Thiamine (%) | Riboflavin (%) |
|---|---|---|
| Microwave | 101 | 87 |
| Electric Oven | 95 | 81 |

Adapted from Campbell (1954)

that the outer layers of a roast should have received a more severe heat treatment than the inner region. The study included three different oven temperature settings (93, 149 and 204 C) in cooking roasts to a center temperature of 60 C. One roast was cooked in a household model microwave oven and removed at a center temperature of 27 C. After 20 minutes standing time the center temperature was 58 C. As predicted, analysis found a higher thiamine content in the inner region except for the case of the roast cooked conventionally in a 149 C oven. The researchers conjectured that the higher thiamine in the outer layers could have been owing to the flow of tissue fluids from the center region (Table 7.13).

Kylen et al (1964) found no significant difference in the thiamine content of pork roasts cooked by microwave or conventional means. There was, however, a significantly higher percentage of thiamine in the drippings that could probably be accounted for by the shorter exposure time that the drippings were exposed in the microwave case compared to the much longer time of exposure in conventional cooking. The same results were evident in the drippings of beef roasts, however the reverse was true for the thiamine retention in the cooked meat. In the cases of beef and ham loaves, there was no significant difference in the thiamine content of the meat (Table 7.14).

Reasons for the differences between the results of Lushbough et al (1962) and Kylen et al (1964) include: roast size, Lushbough used 3.6 to 5.0 kg roasts compared to Kylen et al, 1.55 to 1.86 kg; temperature out of the oven, 27 C compared to 56 to 57 C; final roast temperature, 58 C compared to 67 to 76 C; microwave oven power, consumer oven compared to a commercial oven; cooking time, 45

TABLE 7.13
THIAMINE RETENTION IN BEEF ROASTS COOKED BY MICROWAVE
AND CONVENTIONAL METHODS

| Method | Oven Temperature (C) | Thiamine (%) |
|---|---|---|
| Conventional oven | | |
| Inner portion | 93 | 88 |
| Outer portion | 93 | 77 |
| Inner portion | 149 | 88 |
| Outer portion | 149 | 102 |
| Inner portion | 204 | 67 |
| Outer portion | 204 | 60 |
| Electronic oven | | |
| Inner portion | | 86 |
| Outer portion | | 67 |

Adapted from Lushbough et al (1962)

minutes compared to 20 to 25 minutes. All of these factors translate into milder cooking conditions in the Lushbough et al study. The much longer cooking time in the Lushbough et al study was necessary because of the much larger roasts and the lower microwave power. Standing time was common for both studies. Lushbough's objective was a 60 C roast by both cooking methods. Kylen et al removed the roasts from the oven at just below 60 C and the roast center temperature coasted to 67 to 76 C. The Kylen et al procedure was by all measures much more severe and the analytical data bears this out.

It is interesting to draw a parallel between this work and that of Benjamin Thompson (Count Rumford) in the early 1800s (Brown, 1969). The customary method for cooking roasts was over an open fire and the usual result was much overcooked and inedible meat. Benjamin Thompson designed a roasting oven so that the roast was exposed to a hot atmosphere instead of flames. He not only determined the yield of cooked meat in both cases, but also the quality. In his own words, "To prevent all deception, the persons employed in roasting them were not informed of the principal design of the experiment. When these pieces of roasted meat came from the fire they were carefully weighed; when it appeared that the piece which had been roasted in the roaster was heavier than the other by a difference which was equal to 6% or, 6 lb in 100. But this even is not all; nor is it the most important result of the experiment. These two legs of mutton were brought upon table at the same time and a large and perfectly unprejudiced company was assembled to eat them. They were both declared to be very good; but a decided preference was unanimously given to that which had been roasted in the roaster, it was much more juicy, and was thought to be better tasted. They were both eaten up, and nothing remaining of either of them that was eatable. Their fragments, which had been carefully preserved, being now collected and placed in their separate dishes, it was a comparison

TABLE 7.14
THE EFFECT OF COOKING METHOD ON THIAMINE RETENTION IN MEATS

| Products and Method | Cooked meat (%) | Drippings (%) | Total (%) |
|---|---|---|---|
| Beef roasts I | | | |
| Gas oven | 80 | 2 | 81 |
| Microwave | 58 | 13 | 70 |
| Beef roasts II | | | |
| Electric oven | 86 | 1 | 86 |
| Microwave | 67 | 14 | 80 |
| Pork roasts | | | |
| Electric oven | 61 | 19 | 80 |
| Microwave | 60 | 31 | 91 |
| Beef loaves | | | |
| Gas oven | 76 | | |
| Microwave | 80 | | |
| Ham loaves | | | |
| Gas oven | 91 | | |
| Microwave | 87 | | |

Adapted from Kylen et al (1964)

of these fragments which afforded the most striking proof of the relative merit of these two methods of roasting meat, in respect to the economy of food. Of the leg of mutton which had been roasted in the roaster, hardly anything visible remained except bare bones; while a considerable heap was formed of scraps not eatable which remained of that roasted on the spit."

When properly carried out microwave roasting is a less severe cooking method than conventional oven roasting.

Thomas et al (1982) determined the effect of cooking flake-cut, formed pork roasts on the vitamins thiamine and riboflavin. The flake-cut process represents a method studied by the US Army Natick Laboratories to use carcass meats more economically by forming steaks and roasts from less tender cuts of meat. Because such roasts do not have a normal musculature, they are structurally much weaker, and all cooking, therefore, was carried out from the frozen state. The roasts were formed under pressure (3447 kPa) into uniform cylinders, approximately 10 to 11 cm in diameter and 24 cm long. They weighed 2.0 to 2.5 kg. Roasting was carried out by three methods: a conventional electric oven set at 163 C, a convection oven set at 135 C and a microwave oven operating at 2450 MHz with 300 watts of power. All were cooked to an internal temperature of 77 C. The ends (4 to 5 cm) of the microwave roasts were covered with aluminum foil cuffs for a large part of the cooking cycle to prevent excessive overcooking in this area. The thiamine content of the cooked roasts was similar; i.e., there was no significant difference; whereas the riboflavin content was significantly less for microwave cooking.

Lee et al (1981) found no significant difference in the thiamine content of convection oven cooked chicken that was held for 90 minutes on a steam table, frozen, thawed for 24 hours in a refrigerator and reheated 4 minutes in a microwave oven or 20 minutes in a conventional oven. Most of the loss (33%) occurred during the initial cooking.

Bowers et al (1974a) studied thiamine and riboflavin retention in cooked; cooked and reheated; cooked and frozen; cooked, frozen and reheated turkey muscle using conventional gas oven and microwave oven cooking. They found no significant difference in thiamine content.

Bowers et al (1974b) determined the vitamin B6 content of pork loin cooked by microwave and conventional means and found conventional cooking resulted in significantly greater retention of this vitamin. They added that the differences were small and that variation among animals was greater than between oven types.

### Effect of reheating methods on precooked foods

Causey and Fenton (1951a) compared reheating methods on ascorbic acid retention in several precooked frozen vegetables. The vegetables were precooked by conventionally accepted procedures and cooled immediately and quickly, packaged in 100 gram portions, frozen at −12 F and stored at 0 F for up to 5½ months. Reheating was carried out by five different methods: a Maxson Whirlwind oven (an early forced convection oven) at 300 F with the vegetables on a Pyrex™ plate, unwrapped; in a household electric oven at 400F, also on a Pyrex™ plate, unwrapped; in a Pyrex™ double boiler; immersion in boiling water in a pliofilm bag; and in a microwave oven in a pliofilm bag on a Pyrex™ plate. Details about the microwave oven were not provided other than it was manufactured by Raytheon Company. The only oven available at that time was the 2 kW, 2450 MHz oven (see Chapter One). Their results are shown in Table 7.15.

TABLE 7.15
AVERAGE RETENTION OF ASCORBIC ACID IN FROZEN COOKED VEGETABLES REHEATED BY SEVERAL METHODS

| Method | Time (min) | Cut green Beans | Swiss Chard | Broccoli |
| --- | --- | --- | --- | --- |
| Maxson oven | 16.5 | 71 | 66 | 91 |
| Household oven | 42 | 75 | 64 | 74 |
| Double boiler | 51 | 59 | 79 | 79 |
| Boiling water | 18 | 76 | — | 82 |
| Microwave | 5 | 67 | 85 | 85 |

Adapted from Causey and Fenton (1951a)

Causey and Fenton (1951b) also studied the effect of three reheating methods on the thiamine content of several meat dishes. The differences were not significant for creamed chicken and paprika chicken but were for spaghetti and meat balls and ham patties.

Stevens and Fenton (1951) compared the effect of microwave and stewpan cookery on the retention of vitamins in frozen peas. Portions weighing 300 grams were cooked in the microwave oven on a Pyrex™ plate or in 30 grams of boiling water on a range top. The experiment was replicated eight times and the results are presented in Table 7.16. There was no significant difference in weight loss or palatability owing to cooking method.

Kahn and Livingston (1970) reported on the effect of reheating methods such as water immersion, forced convection, infra red and microwave on nutrient retention. They compared the thiamine content of beef stew, chicken a la king, shrimp neuburg, and peas in cream sauce, freshly prepared and held at 180 F for one, two and three hours with the frozen products reheated in boiling water and by microwaves or infra red radiation. The results are shown in Table 7.17. The average losses of

TABLE 7.16
RETENTION OF VITAMINS IN COMMERCIALLY FROZEN PEAS COOKED BY MICROWAVE AND CONVENTIONAL MEANS

| Method | Ascorbic Acid (%) | Thiamine (%) | Riboflavin (%) |
| --- | --- | --- | --- |
| Microwave | 83 | 95 | 98 |
| Stewpan | 86 | 97 | 100 |

Adapted from Stevens and Fenton (1951)

TABLE 7.17
THIAMINE RETENTION OF FRESH AND FROZEN FOOD PRODUCTS BY DIFFERENT REHEATING METHODS

| Method | Thiamine Retention (%) |
| --- | --- |
| Freshly prepared | 100.0 |
| Frozen & Microwave heated | 93.5 |
| Frozen & Infrared heated | 90.4 |
| Frozen & Boiling water heated | 86.0 |
| Fresh & held hot 1 hour | 78.2 |
| Fresh & held hot 2 hours | 73.9 |
| Fresh & held hot 3 hours | 67.4 |

Adapted from Kahn and Livingston (1970)

thiamine were 6.5% for microwave heating, 9.4% for infrared heating and 14% for hot water immersion. By way of contrast, holding the freshly prepared products at 180 F in a bain marie for 1, 2 and 3 hours resulted in thiamine losses of 22, 26, and 33%. All of the reheat methods used were obviously better than fresh preparation and hot holding for one hour.

Preparing foods and holding them hot during plating for patient foodservice has been the practice in many health care institutions. Even when prepared frozen foods in bulk containers such as one-half and full size steam table pans are used, the usual method of heating these foods to serving temperature results in a situation that is not conducive to nutrient retention; i.e., a whole oven full of ready-to-serve food that must be maintained hot until served, all the while undergoing significant degradation of labile vitamins.

Kahn and Livingston (1970) suggested that rapid heating methods may be important in terms of nutrients consumed under marginal feeding conditions as in institutions operating under extreme budget limitations. Their study on the effect of heating methods on thiamine retention indicated that the difference between freshly prepared food held hot and microwave heated prepared foods was equal to 18.4% of the RDA for four to six year olds and 14.6% of the RDA for six to eight years olds and 18 to 75 year old women.

Livingston et al (1973) pointed out that the average American consumes more than one out of four meals away from home (closer to one out of three today) and therefore this meal should possess a caloric density close to the Recommended Dietary Allowance. It is relevant then that commercial foodservice practice should be geared to optimize the retention of naturally occurring nutrients in the food served. In this same article, Livingston et al discussed modern feeding systems in which food preparation and service are separated. Prepared foods may be held hot, chilled or frozen and transported to the serving facility. Holding foods hot is typical of many foodservice operations and the holding time may vary from a few minutes to several hours. In some cases insulated containers are used and in others the foods are kept warm in steam tables.

Much lower losses were found when foods were held chilled for three days than when held hot for three hours. Some data (Kossovitsas et al (1971) indicated that chilled holding resulted in losses not much different from frozen holding for ascorbic acid in broccoli in cream sauce and thiamine in chicken a la king.

Lachance et al (1973) noted that our knowledge of the loss in nutritional values of food during processing, storage and reheating is pitifully poor. They proposed that the most efficient and least cost method would be nutrification of processed food items to assure expected levels of nutrients. It is distinctly possible that this approach will be rendered unnecessary by advances in microwave heating practices.

Ang et al (1975) reported on the effects of heating methods on vitamin retention in six fresh and frozen prepared food products. The foods were packed, 4 to 5 pounds of product in aluminum pans or polyester pouches, and frozen. Heating methods included forced convection, high pressure steam, infra red and microwaves. After heating by these methods the foods were held at 180 F for 30 minutes to simulate normal delay between heating and consumption. In general, freshly prepared products were of the highest nutritive quality and holding warm for up to 3 hours reduced thiamine and ascorbic acid appreciably. Microwave heating required shorter heating times and generally resulted in the highest retention of vitamins. Both microwave and infra red heating retained as much as or more nutrients than fresh preparation and hot holding for up to 1.5 hours. Convection oven heating took the longest time, but retentions were somewhat comparable to microwave and infra red heating.

Snyder (1976) reported on an experiment in microwave heating of vegetable portions by the cafeteria customer. Vegetables were prepared by foodservice personnel in the best manner to the desired degree of doneness, quickly chilled, portioned into serving dishes, overwrapped with film and stored at 34–36 F. At meal time the dishes of vegetables were displayed on ice on the serving line for the customer to take and heat in microwave ovens in the dining area. The results indicated high consumer acceptance and an increase in vegetable sales of 36% over normal vegetable sales from the steam table. Although nutritional data was not taken, the comment was made that nutrient loss by this method would be as low as it is possible to achieve by any food preparation method known today.

Clearly, methods that involve a minimum of food handling before service will provide the consumer with the highest percentage of labile nutrients. Thus, for example, hospital meals plated from refrigerated (chilled) components and microwave heated just before service should result in the patient ingesting a higher percentage of nutrients than if the meal were plated hot and maintained hot until served. The duration of the hot holding period is the critical factor.

Livingston (1979), in a review of research to date (1979) concluded that:

1. The effect of microwave heating on nutrients depends on factors such as the size and shape of the food product, thickness, quality and techniques used in preparation such as stirring and holding;
2. More water soluble nutrients are retained during microwave heating of frozen vegetables by avoiding the leaching effect of a water cook;
3. When microwave heating times are short, greater retention of the heat labile nutrients is the norm. But when heating times are comparable to other methods, the retention of nutrients will be comparable;
4. Microwave heating of frozen prepared foods is equal to or better than conventional methods of preparation and hot holding for one hour or more.

Brittin and Trevino, (1980) conducted a study to determine the acceptability of microwave and conventionally baked potatoes. Microwave baking gave significantly greater cooking losses and had lower sensory scores than conventionally baked potatoes. However, consumers polled at supermarket demonstrations indicated no significant difference in preference or acceptability of potatoes cooked by the two methods.

## Moisture

Most studies reported in the literature indicate a significantly greater moisture loss in foods cooked by microwave energy than for other cooking methods. It is easy to understand why such results were obtained. Not only is microwave cooking more rapid, but the evaporative losses are also greater because of the lower ambient temperatures in microwave ovens. Thus there tends to be a positive pressure gradient toward the food surface, which coupled with the low vapor pressure surrounding the product, leads to greater moisture removal. When cooking food uncovered in a microwave oven, it is not unusual to find considerable condensate on the floor and walls of the oven. In recognition of this, most microwave ovens are designed so that warm air, which passes through the cooling fins of the magnetron to maintain its temperature within operating specification, is vented through the oven in such a manner that some of the excess moisture from cooking is carried away.

Stevens and Fenton (1951) found no significant difference in weight loss for peas cooked by microwaves or stewpan cookery. They did point out that palatability was affected by slight over or undercooking (by 30 seconds) though no change in ascorbic acid occurred.

Apgar et al (1959) Compared microwave and conventional cooking of pork roasts, patties and chops for fat, moisture, thiamine, palatability, and color changes. Both institutional and consumer ovens were used, but the institutional oven was set to provide the same power as the low setting on the consumer oven; i.e., 450 watts. The consumer oven also had a broiler unit and this technique was also used as a variable. All products were cooked to an internal temperature of 87.8 C (190 F). The results indicated no significant differences in cooking loss due to cooking method for patties and roasts. The losses were significantly less for microwave cooking of pork chops (Table 7.18).

Marshall (1960) compared microwave and conventional cooking of top rounds of beef to 80 C and found significantly more moisture loss from microwave cooking. Korschgen and Baldwin (1978) on the other hand found no significant difference in moisture loss when beef rounds were cooked by moist heat microwave and conventional oven braising. This is to be expected when comparing dry with moist heat methods. In the latter case, the moisture losses are high but equivalent because the roasts are cooked to near 100 C. In the former case the microwave method gave the higher moisture losses because (1) the roasts were thin at one end and this end became so hard (overcooked) that it was unpalatable; and (2) the roasts were cooked well done (176 F) which contributed a great deal to the overcooking at the thin ends and elsewhere. Marshall also reported trimming losses (15.8%), that along with drip (13.8%) and evaporation (28.4%), resulted in 39.4% acceptable meat for microwave roasting while conventional roasting yielded 65.3% acceptable meat.

Headley and Jacobson (1960) compared the cooking losses of boned and rolled lamb roasts cooked by microwave and conventional means to 180 F and observed some 8% greater loss for microwave cooking. Roasts were cooked to 150 F in the

TABLE 7.18
THE EFFECT OF COOKING METHOD ON THE THIAMINE CONTENT
OF PORK PRODUCTS

|  | Conventional | Microwave | | Institutional Oven |
|---|---|---|---|---|
|  |  | Consumer Oven | | |
|  |  | With Browning | W/O Browning |  |
| Pork Patties |  |  |  |  |
| Moisture (%) | 47.78 | 47.90 | 46.62 | 47.58 |
| Thiamine (%) dry, fat free | 54.00 | 54.05 | 55.48 | 53.88 |
| Pork Roasts |  |  |  |  |
| Moisture (%) | 55.5 | 52.92 | 51.98 | 53.20 |
| Thiamine (mg/100 gms) | 1.58 | 1.63 | 1.43 | 1.51 |
| Pork Chops |  |  |  |  |
| Moisture (%) | 54.0 | 57.5 | 58.68 | 57.80 |
| Thiamine (mg/100 gms) | 1.662 | 1.822 | 2.060 | 2.080 |

From Apgar et al (1959)

microwave oven and allowed to coast up to 180 F for comparison with the roasts cooked conventionally to 180 F. The higher losses by evaporation for the microwave cooked roasts (27% compared to 19%) could have been due at least in part to this coasting period. Drip losses were similar for both methods (15%—microwave; 17%—conventional).

Phillips (1960) found no significant difference in weight of chicken cooked by microwave or conventional methods. Wing and Alexander (1972) found that chicken breast lost more moisture when microwave cooked than conventionally cooked.

Kylen et al (1964) found that moisture losses in beef and pork roasts and ham loaves were significantly higher for microwave cooking than conventional cooking, but identical for beef loaves. The latter result could be explained by the lack of musculature; that is, by neither method of cooking would there be expression of juices by the mechanical forces of muscle contraction (Table 7.19).

Baldwin et al (1976) cited results of low power (492 watts) and high power (1054 watts) microwave and conventional (163 C) cooking of beef, pork and lamb roasts to 70 C. It should be noted that each oven was operated on a 50% duty cycle so that the average power level over the cooking cycle was half of the power levels given; i.e., 246 and 527 watts, respectively. It should be noted further that the higher powered oven was operated on a cycle of 6 minutes on and 6 minutes off, while the lower powered oven was operated on a cycle of 3 minutes on and 3 minutes off. Essentially what this means is the 1054 watt oven is cooking for 6 minutes at full power then is off for six minutes, while the 492 watt oven is cooking for 3 minutes

TABLE 7.19
THE EFFECT OF COOKING METHOD ON COOKING LOSSES

| Method | Average Cooking Loss (%) | | |
|---|---|---|---|
| | Evaporation | Drippings | Total |
| Beef Roasts | | | |
|   Gas oven | 10.2 | 7.4 | 17.5 |
|   Microwave | 20.0** | 18.8** | 38.8** |
|   Electric oven | 12.8 | 8.6 | 21.4 |
|   Microwave | 16.4 | 13.3** | 29.3* |
| Pork roasts | | | |
|   Electric oven | 21.9 | 12.3 | 34.1 |
|   Microwave | 20.2 | 17.1** | 37.3 |
| Beef loaves | | | |
|   Gas oven | 15.2 | 9.1 | 24.3 |
|   Microwave | 18.0** | 8.9 | 26.9* |
| Ham loaves | | | |
|   Gas oven | 13.0 | 5.2 | 18.2 |
|   Microwave | 22.8** | 5.6 | 28.3** |

Adapted from Kylen et al (1964)
\* Significantly higher at the 5% level for microwave cooking
\*\* Significantly higher at the 1% level for microwave cooking

at full power and then is off for 3 minutes. Although cooking times were not given and since the roasts were similar in size, roasts cooked in the 1054 watt oven should have been cooked in substantially less time than in the 492 watt oven. Considering the difference in heating rates the differences in the moisture, fat and protein content of the cooked roasts are remarkably close. Moisture of cooked beef roasts was significantly lower for microwave cooking (52.9 and 53.5%, respectively, for high and low power compared to 59.9% for conventional cooking). There was no significant difference in the moisture content of pork or lamb roasts cooked by either method. Evaporative losses, i.e., moisture content of the drippings, for all roasts were significantly higher for the microwave methods. The protein content found in the drippings of roasts cooked by the microwave method was significantly lower than in the drippings of conventionally cooked roasts. The results are summarized in Table 7.20.

Korschgen et al (1976) in a study of the effect of cooking method on quality factors of beef, pork, and lamb roasts also reported on cooking losses. Their study was carried out using the longissimus muscle of beef and pork, and deboned leg of lamb. Microwave cooking at two power levels, 1054 and 492 watts, was used; the higher powered oven was cycled on and off in 3-minute intervals, while the lower powered oven was cycled at 6-minute intervals. Conventional cooking was carried out at 163 C in a gas-fired oven. The roasts were removed from the microwave

TABLE 7.20
THE EFFECT OF COOKING METHODS ON THE MOISTURE, FAT, PROTEIN
AND VITAMIN CONTENT OF MEAT ROASTS

|  | Microwave oven | | Conventional oven |
|---|---|---|---|
|  | 1054 W | 492 W | (163 C) |
| **Beef** | | | |
| Moisture | 52.9 | 53.5 | 59.9 |
| Fat | 15.1 | 14.7 | 11.3 |
| Protein | 30.8 | 31.5 | 28.5 |
| Thiamine | 61.0 | 49.0 | 69.0 |
| Riboflavin | 98.0 | 83.0 | 99.0 |
| Niacin | 94.0 | 86.0 | 104.0 |
| **Pork** | | | |
| Moisture | 62.1 | 60.4 | 62.1 |
| Fat | 4.5 | 5.3 | 4.0 |
| Protein | 33.2 | 33.6 | 33.6 |
| Thiamine | 73 | 67 | 72 |
| Riboflavin | 81 | 82 | 96 |
| Niacin | 87 | 79 | 101 |
| **Lamb** | | | |
| Moisture | 59.1 | 60.6 | 60.7 |
| Fat | 7.5 | 7.2 | 8.1 |
| Protein | 32.3 | 31.5 | 27.3 |
| Thiamine | 52 | 49 | 52 |
| Riboflavin | 88 | 73 | 98 |
| Niacin | 71 | 64 | 86 |

Adapted from Baldwin et al (1976)

ovens at predetermined internal temperatures in recognition of post oven temperature rise and allowed to coast up to 70 C. The cooking losses are summarized in Table 7.21. It is clear that the losses for beef are greater for microwave cooking regardless of oven power, but not significantly so for pork or lamb.

Baldwin et al (1976) noted that the higher moisture content in the drippings of roasts cooked at the lower power level (492 watts) are probably owing to the lower ambient temperature in the oven permitting a greater percentage of the liberated moisture to condense. The substantially longer conventional cooking time and the higher oven temperature (163 C) allows a greater moisture loss. Why this is not so for lamb roasts may be related to the much higher fat content in the drippings.

Drew et al (1980) studied the effect of cooking at variable power levels on the quality of top round beef roasts. They pointed out this feature of today's microwave ovens was being promoted as the preferred procedure for meat cookery in terms of improved quality of the cooked meat. Beef roasts weighing about a kilogram (2.2

TABLE 7.21
EFFECT OF COOKING METHOD ON AVERAGE COOKING LOSSES IN BEEF, PORK AND LAMB

| Method | Cooking Losses (%) | | |
|---|---|---|---|
| | Evaporation | Drip | Total |
| Beef | | | |
| MW 1054 Watts | 24.4 | 6.4 | 29 |
| MW 492 Watts | 20.0 | 9.7 | 29.8 |
| Conventional | 16.8 | 3.5 | 20.2 |
| Pork | | | |
| MW 1054 Watts | 22.3 | 11.7 | 34.0 |
| MW 492 Watts | 17.8 | 17.4 | 35.2 |
| Conventional | 29.9 | 5.0 | 34.9 |
| Lamb | | | |
| MW 1054 Watts | 28.9 | 6.4 | 35.2 |
| MW 492 Watts | 23.5 | 9.3 | 32.8 |
| Conventional | 26.8 | 4.4 | 31.1 |

Adapted from Korschgen et al (1976)

lb) and of similar geometry (ca 17 × 6.5 cm) were cooked from the frozen (−18 C) as well as the thawed (5 C.) state. Microwave cooking was carried out at 553 watts and 237 watts, the high and simmer settings of the oven). Conventional cooking was carried out in an electric oven at 163 C. It should be noted that all consumer microwave ovens do not operate on the same time base; that is, the ovens operate at full power regardless of power setting. The time base may vary from two to 60 seconds, thus 50 per cent power could be 1 second off and 1 second on, or 30 seconds on and 30 seconds off.

## SUMMARY

After a comprehensive review of the changes in nutrient and chemical composition of foods heated by microwaves, Lorenz (1976) concluded that when foods are heated to the same final temperature, those prepared by the microwave method are as nutritious as those prepared by conventional methods.

Most of the changes of significance are those owing to preparation method. The little data available in the literature on effect of reheating method tends to favor microwave reheating as least harmful to vitamins.

In both cooking and reheating, the data suggest that microwave heating when properly carried out will provide consistently better nutrition than any other method of cooking or reheating. In the case of fruits and vegetables this is accomplished by essentially waterless cooking with microwaves. Properly prepared and preserved foods will reheat with a great deal of convenience so that such foods will provide optimum nutrition. The combination of convenience and good nutrition provides a benefit that no other food preparation method can offer.

## REFERENCES

Ang, C.Y.W., Chang, C.M., Frey, A.E. and Livingston, G.E. (1975). Effects of heating methods on vitamin retention in six fresh or frozen prepared food products. J. Food Sci. 40 (5) 997–1003.

Apgar, J., Cox, N., Downey, I. and Fenton, F. (1959). Cooking pork electronically. J. Amer. Dietet. Assoc. 35 (12) 1260.

Armbruster, G. (1978). Comparison of reduced ascorbic acid content of microwave and conventional fruits and vegetables. Final report. Thermador/Waste King Pamphlet, MHS-1.

Baldwin, R.E., Korschgen, B.M., Russell, M.S. and Mabesa, L. (1976). Proximate analysis, free amino acid, vitamin and mineral content of microwave cooked meat. J. Food Sci. 41:762–765.

Brittin, H.C. and Trevino, J.E. (1980). Acceptability of microwave and conventionally baked potatoes. J. Food Sci. 45:1425–1427.

Bowers, J.A., Fryer, B.A. and Engler, P.P. (1974a). Vitamin B6 in turkey breast muscles cooked in microwave and conventional ovens. Poultry Sci. 53, 844.

Bowers, J.A., Fryer, B.A. and Engler, P.P. (1974b). Vitamin B6 in pork muscle cooked in microwave and conventional ovens. J. Food Sci. 39 (2) 426–427.

Brown, S.C. (1969). The Collected Works of Count Rumford, Vol III. Harvard Univ. Press, Cambridge, Massachusetts.

Campbell, C. (1954). Personal communication.

Campbell, C. and Gretchell, E. (1957). Personal communication.

Campbell, C., Lin, T.Y. and Proctor, B.E. (1958). Microwave vs conventional cooking. I. Reduced and total ascorbic acid in vegetables. J. Am. Dietet. Assn. 34 (4) 365–370.

Causey, K. and Fenton, F. (1951a). Effect of reheating on palatability, nutritive value, and bacterial count of frozen cooked foods. I. Vegetables. J.A.D.A. 27 (5) 390–395.

Causey, K. and Fenton, F. (1951b). Effect of reheating on palatability, nutritive value, and bacterial count of frozen cooked foods. II. Meat dishes. J.A.D.A. 27 (6) 491–495.

Chapman, V.J., Putz, J.O., Gilpin, G.L., Sweeney, J.P. and Eisen, J.N. (1960). Electronic cooking of fresh and frozen broccoli. J. Home Econ. 52:161.

Cross, G.A. and Fung, D.Y.C. (1982). The effect of microwaves on nutrient value of foods. CRC Rev. Food Sci. & Nutr. 16 (4) 355–381.

Dietrick, W.C., Huxell, C.C. and Guadagni, D.G. (1970). Comparison of microwave, conventional and combination blanching of Brussels sprouts for frozen storage. Food Technol. 24 (5) 105–108, 109.

Drew, F., Rhee, K.S. and Carpenter, Z.L. (1980). Cooking at variable microwave power levels. J. Amer. Dietet. Assn. 77:455–459.

Eheart, M.S. and Gott, C. (1964). Conventional and microwave cooking of vegetables. J. Am. Diet Assoc. 44:116.

Eheart, M.S. and Gott, C. (1965). Chlorophyll, ascorbic acid, and pH changes in green vegetables cooked by stir fry, microwave and conventional methods and a comparison of chlorophyll methods. Food Technol. *19* (5) 185–188.

Gordon, J. and Noble, I. (1959). Comparison of electronic and conventional cooking of vegetables. J. Am. Dietet. Assn. *35* (3).

Headley, M.E. and Jacobson, M. (1960). Electronic and conventional cookery of lamb roasts. J.A.D.A. *36* (4) 337–340.

Kahn, L.N. and Livingston, G.E. (1970). Effect of heating methods on thiamine retention in fresh or frozen prepared foods. J. Food Sci. *35*:349–351.

Korschgen, B.M., Baldwin, R.E. and Snider, S. (1976). Quality factors in beef, pork, and lamb cooked by microwaves. J. Amer. Dietet. Assoc. *69*:635–670.

Korschgen, B.M. and Baldwin, R.E. (1978). Sensory qualities, cooking losses, shear values, and B-vitamins of beef roasts cooked by slow-heat. Home Econ. Res. J. *7*(2) 116.

Kossovitsas, C., Navab, M., Chang, C.M. and Livingston, G.E. (1973). A comparison of chilled-holding versus frozen storage on quality and wholesomeness of some prepared foods. J. Food Sci. *38*:901–902.

Kylen, A.M., Charles, V.R., McGrath, B.H., Schleter, J.M., West, L.C. and Van Duyne, F.O. (1961). Microwave cooking of vegetables. J. Amer. Dietet. Assoc. *39*:321–326.

Kylen, A.M., McGrath, B.H., Hallmark, E.L. and Van Duyne, F.O. (1964). Microwave and conventional cooking of meat. J. Amer. Dietet. Assoc. *45*:139–145.

Lachance, P.A., Ranadive, A.S. and Matas, J. (1973). Effects of processing, storage and handling on nutrient retention in foods: effects of reheating convenience foods. Food Technol. *27* (1) 36–38.

Lanier, J.J. and Sistrunk, W.A. (1979). Influence of cooking method on quality attributes and vitamin content of sweet potatoes. J. Food Sci. *44*:374–376, 380.

Lee, F.V., Khan, M.A. and Klein, B.P. (1981). Effect of preparation and service on the thiamin content of oven-baked chicken.

Livingston, G.E., Ang, C.Y.W. and Chang, C.M. (1973). Effects of processing, storage and handling on nutrient retention in foods: effects of food service handling. Food Technol. *27* (1) 28–30, 32, 34.

Livingston, G.E. (1979). Nutrition aspects of microwave cooking. Microwave Energy Applications Newsl. *12* (5) 4, 6–9.

Lorenz, K. (1976). Microwave heating of foods-changes in nutrient and chemical composition. CRC Rev. Food Sci. & Nutr. *7* (4) 339–370.

Lushbough, C.H., Heller, B.S., Weir, E. and Schweigert, B.S. (1962). Thiamine retention in meats after various heat treatments. J. Am. Diet Assoc. *40*, (1) 35–38.

Marshall, N. (1960). Electronic cookery of top round of beef. J. Home Econ. *52*, January, 31–34.

Phillips, L., Delaney, I. and Mangel, M. (1960). Electronic cooking of chicken. J.A.D.A. *37* (5) 462–465.

Proctor, B.E. and Goldblith, S.A. (1948). Radar energy for rapid cooking and blanching and its effect on vitamin content. Food Technol. *2*:95-104.

Semling, (1984) Examining food labelling and advertising policy. Food Proc. *45* (13) 12–13.

Snyder, O.P. Jr. (1976). Increasing the quality of vegetables in foodservice operations. Microwave Energy Appl. Newsl. *9* (2) 3–7.

Stevens, H.B. and Fenton, F. (1951). Dielectric vs stewpan cooking. Comparison of palatability and vitamin retention in frozen peas. J.A.D.A. *27* (1) 32–35.

Thomas, M.H., Brenner, S., Eaton, A. and Craig, V. (1949). Effect of electronic cooking on nutritive value of foods. J. Am. Dietet. Assn. *25* (1) 39–44.

Thomas, M.H., Decareau, R.V. and Atwood, B.M. (1982). Thiamin and riboflavin content of flake-cut formed pork roasts. J. Microwave Power *17* (1) 83–87.

Wing, R.W. and Alexander, J.C. (1972). Effect of microwave heating on vitamin B6 retention in chicken. J.A.D.A. *61*:661.

# CHAPTER EIGHT

# MICROBIOLOGICAL CONSIDERATIONS

Some degree of shelf-stability in foods is essential to increase the distribution range for a manufactured product. Shelf-stable foods currently represented by canned (including jars) foods, provide the longest shelf-life, though at some sacrifice in quality. There are efforts being made today to improve the quality of such foods by means of a microwave sterilization process, particularly foods in microwavable packaging. Frozen foods provide another solution to this problem. Frozen foods, however, require a large support organization. Success depends on a sequence of steps including preparation, packaging and rapid freezing to preserve quality; controlled temperature distribution; storage and handling in the sales outlet; and finally, proper handling and preparation by the consumer. Failure to carry out any of these steps properly can lead to poor results and a dissatisfied customer.

Chilled foods represent still another category of food for the microwave oven user that has significant potential. These foods, however, have a much shorter shelf-life that is dependent on very tight quality control and distribution procedures to be effective. Microbiological factors play a major role in the life of these products and present the manufacturer with a difficult scheduling program that must not be violated since the public health is at stake. The shelf-life of chilled foods can be extended for some products by modified atmosphere (MA) packaging, an in-package microwave pasteurization step or a combination of MA packaging and microwave pasteurization. MA packaging of fresh pasta adds weeks to its salable life and microwave pasteurization of MA packaged pasta claims to add substantially more shelf-life. This process has become quite popular in Italy and in other European countries. More than 2000 tons per day are being processed. It does not appear to be gaining popularity in the United States. In-package microwave pasteurization of sliced specialty breads is another process practiced to some degree in Europe where the use of preservatives is frowned upon.

Genigeorgis (1988) noted that MA packaging cannot eliminate the risk from microbial hazards, but that the risk could be minimized if food processors

1. used ingredients with low counts of pathogenic bacteria;
2. used thermal processes that kill non-sporeforming pathogens and continuously monitor such processes;
3. take measures to avoid post process contamination before packaging;
4. store product as near as possible to 0 C.;

5. develop formulations that provide an environment not conducive to growth of pathogens in case of temperature abuse; for example, acidification, reduced water activity, use of antimicrobials and microbial competitors;

6. use a Hazard Analysis and Critical Control Point (HACCP) system throughout the process, distribution and storage.

The effect of microwave energy on microorganisms in food was reviewed in some detail by Fung and Cunningham (1980). The mechanism by which microorganisms are destroyed is generally conceded to be thermal; that is, by heat generated in the food by microwave energy absorption. Some researchers (Fleming, 1944; Nyrop, 1946; Zarotschenzeff, 1944; Olsen et al, 1966; Gray, 1970a,b,c,) argued that there is an athermal effect, and much publicity has been given to such speculation. The implication is made that foods can be sterilized with little or no heat. Efforts by others (Brown et al, 1947; Brown and Morrison, 1954; Proctor and Goldblith, 1951; Ingram and Page, 1953; and Wang and Goldblith, 1967) to confirm these results were all negative. Numerous other studies (Dessel et al, 1960; Lacey et al, 1965; Olsen et al, 1966, Baldwin et al, 1968; Delaney et al, 1968; Ockerman et al, 1977) seemed to indicate that microwave energy was less effective in reducing the microbial population of foods than conventional methods of heating. Many of the negative results can be explained by uneven heating and evaporative cooling of uncovered foods that could have been avoided by simple expedients such as heating the foods covered or at a reduced heating rate. Olsen et al (1966) observed surface cooling when microwave cooking. Overwrapping with plastic film prevented evaporative cooling, and when inoculated frankfurters were plastic wrapped and microwave heated, there were no surviving organisms.

## MICROBIOLOGY OF COOK/CHILL FOODSERVICE

Foods served to hospital patients should be microbiologically safe whether or not the food is prepared daily from fresh ingredients or prepared and chilled for service at a later time. Foodborne illnesses do occur in hospitals. Many have been recorded and probably many more have not. Contamination by a food handler can occur whether the food is prepared in the hospital kitchen or by an outside supplier. Even if a food handler is not implicated, if bacteria are present they will multiply rapidly if time and temperature conditions are favorable.

Analysis of surveillance data for the period 1968 to 1977 by the Centers for Disease Control indicated that meat and poultry products were implicated in more than 50% of the outbreaks of foodborne diseases. Some 69% of the outbreaks were in foodservice establishments. The most frequent causes were: improper cooling of cooked foods (48%), foods prepared a day or more before serving (34%), inadequate cooking or thermal processing (27%), infected person touching cooked foods (23%) and inadequate heating of cooked and chilled foods (20%) (Bryan, 1980).

Perhaps one of the most well-known examples of a cook/chill foodservice system was that used by the Naaka Hospital in Sweden. The system was described in some detail by Bjorkman and Delphin (1966). Although the Naaka system did not use microwave heating, a brief discussion of the system is germane to this text because it led to the growing interest in chill foodservice systems, many of which do use microwave ovens.

The two major benefits claimed for the Naaka system, aside from the prospects of increased productivity and other labor-related advantages, was the safe microbiological condition of the prepared food, and the good quality without the need to resort to freezing. Basically, food is prepared in the usual manner, being sure that a food temperature of at least 176 F (80 C) is reached. It is transferred hot to plastic bags, usually five portions of a main dish per bag. The air is removed and the bags sealed, placed in boiling water for three minutes, then cooled in running water to 37 F (2.8 C). After they are dried, the bags are stored at 37 F (2.8 C) or less. To use, bags are placed in boiling water for 30 minutes, opened and the contents plated. Generally, only the main dishes are processed in this manner. Potatoes and vegetables are cooked just before serving.

Microbiological safety is assured by cooking all parts of the food to 176 F (80 C) and then, after packaging, pasteurizing in boiling water to kill any bacteria that may have entered the package. Although some bacterial spores could have survived this treatment, the rapid chilling of the packaged foods would have hindered their growth. At the time this paper was presented over five years had passed during which some five million portions had been prepared and served, and over 100,000 microbiological analyses had been performed. The results demonstrated that the method gave a completely satisfactory product. The National Veterinary Board of Sweden, after studying the results, stated that, "it finds that this method of preparation meets the highest hygienic requirements in a satisfactory manner."

A variation of the Naaka system was developed in the United States in the late 1960s by McGuckian (1971) and was called the A.G.S. System after the initial letters of Anderson, Greenville and Spartenburg, South Carolina where three hospitals and the Duke Foundation sponsored a study to develop an improved foodservice system. The major difference between the two systems was that in the A.G.S. System the food was partially cooked then individual portions were vacuum packaged in plastic bags and the cooking completed in the bag by immersion in boiling water. According to McGuckian, this procedure captures and retains the flavor of food at its peak of excellence. One other difference between the systems was the 10 to 20 second microwave heating of plated portions just prior to serving thus insuring that the food is served hot.

Kossovitsas et al (1973) studied the effect of pasteurization on *Salmonella* and *Clostridium perfringens* inoculated into prepared foods following the general procedures developed for the Naaka system. The products tested were chicken a la king, codfish in cream sauce, and broccoli in cream sauce. Individual portions were filled into plastic pouches, vacuumized, sealed, pasteurized and chilled. Chilled samples were stored at 35.6 F (2 C) and frozen samples at $-10$ F. ($-23.3$ C). After 15 and 30 days storage the chilled samples, following heating in boiling water, indicated no survivors, whereas Salmonella survived frozen storage and heating to 154 F.(67.7 C). The frozen samples had not been vacuum packaged and pasteurized. Panelists were unable to detect significant differences between frozen and refrigerated samples after 15 days storage, though fresh controls were rated superior to both. However, after 30 days storage, the panel found the refrigerated samples unacceptable.

A major factor that affects bacterial growth is the time a prepared food product remains in the temperature range of 45 to 140 F (6.1 to 60 C). Longree (1972) stated that food preparation systems should use equipment that can cool foods through this temperature range in four hours or less. Some studies have shown that this goal is not always attained.

Bunch et al (1976) studied the acceptability and microbiological characteristics of beef-soy loaves processed according to hospital cook/chill foodservice practice. The loaves were cooked in a convection oven to 140 F (60 C), held at 41 F (5 C) for 24, 48, and 72 hours, portioned onto plates and heated in microwave ovens to 176 F (80 C). The time to chill to 41 F (5 C) in the walk-in refrigerator used in this study was approximately 8 hours. Although the aerobic bacteria population increased during chilled storage, microwave heating decreased the numbers sufficiently so that a conclusion could be drawn that, product handled according to these procedures is in excellent bacteriological condition when it reaches the consumer.

In some hospitals using cook/chill foods for following day service the chilled foods are plated, placed in refrigerator carts, shipped to the appropriate wards and heated in microwave ovens. Many of these hospitals are not equipped with sophisticated means for rapid chilling. Nicholanco and Matthews (1978) carried out a study in one such hospital and found that when freshly prepared beef stew was portioned six quarts to 12 $\times$ 10 $\times$ 4 inch counter pans, the internal and surface temperatures remained above 45 F (7.2 C) for more than nine hours when cooled in a walk-in refrigerator. The refrigerator temperature remained between 42 F (4.4 C) and 50 F (10 C) during these nine hours. Samples of stew were portioned into individual bowls and placed in refrigerated carts for a period of time to simulate distribution, then heated in microwave ovens. Bacteria counts made immediately after cooking, after 3, 5, 7, 9, and 19.5 hours chilled storage, after two hours holding and after microwave heating indicated that the aerobic bacteria count was lowest immediately after cooking and microwave heating and highest during chilled storage. In spite of the relatively slow cooling cycle, aerobic plate counts did not increase appreciably and reheating portions of beef stew to 145 F (63.8 C) in a microwave oven reduced the plate count to the same level as that at completion of cooking (6.6 x $10^4$). US Army and Skylab standards are $1 \times 10^5$ and $1 \times 10^4$, respectively, at point of service.

The arguments being offered in support of cook/chill foodservice systems are compelling: less energy and equipment required, shorter reheat time, no thawing problem, lower costs. Storage at 40 F (4.4 C), which is typical of these systems, does not insure microbiological safety. Mesophilic organisms can grow at this temperature.

Some organisms, such as *Staphylococcus aureus* and *Salmonella typhimurium*, can increase to danger levels without altering the appearance, odor or flavor of prepared foods. To minimize bacteria growth, the minimum internal temperature of cooked foods should be no less than 165 F (73.3 C), and the foods should be chilled to 45 F (7.2 C) in two hours followed by storage at not over 40 F (4.4 C). Storage time should not exceed nine days. Foods should be reheated to 165 F (73.3 C) and served at no less than 150 F (65.5 C). These recommendations by Rowley et al (1972) were the outcome of a study in Army foodservice at Fort Lewis, Washington carried out by the US Army Natick Research and Development Laboratories in the early 1970s. Rapid chilling in the Fort Lewis study was accomplished in a liquid nitrogen cooled box with a capacity of 235 lbs of foods in 10 × 12 × 2 inch pans. The chilling time to 40 F (4.4 C) or less was not over two hours.

A system developed by the W.R. Grace Company, Duncan, South Carolina, for the chilling of mostly sauced type foods and soups is even more rapid than the liquid nitrogen system described above. In this system, the cooked foods are pumped into special two gallon plastic bags, sealed, then agitated in a tumbling device, much like a clothes dryer, while being sprayed with cold water. Called the CAPKOLD System, chilling to less than 40 F (4.4 C) is achieved in about 20 minutes.

## REVIEW OF THE MICROBIOLOGY OF MICROWAVE COOKED AND HEATED FOODS

It is difficult to compare results of the many researchers who have carried out studies on the effect of microwave heating on the microbiological population of food in foodservice situations. Not only are details lacking, but most of the studies were carried out with a less than complete knowledge of microwave heating phenomena. There is essentially no literature available dealing with microwave heating variables on plate counts; e.g., reducing power level and increasing heating time, covering food during heating, shielding, and food geometry to name a few. There are also differences among the many microwave oven designs on the market that contribute to the difficulty.

One of the earliest studies on the effect of microwave reheating on the microbiological quality of frozen prepared foods was carried out at Cornell University. Causey and Fenton (1951b) determined the bacteria count in laboratory produced creamed chicken and rice, paprika chicken and gravy, spaghetti and meat balls and ham patties. Initial counts were less than 2000/gram. Organisms identified were *Bacillus cereus*, *Bacillus megatherium* and *Staphylococcus* strains including some which were coagulase positive. By way of contrast a commercial chicken a la king product at that time averaged 50,000/gram. Reheating to 180 to 185 F (82.2 to 85 C) by several methods including a Maxon Whirlwind (convection) oven, double boiler, boiling water immersion and microwave oven reduced counts to 6/gram. Similar results were obtained with vegetables — green beans, Swiss chard, broccoli, diced carrots, and shredded beets (Causey and Fenton, 1951a). The microwave oven used in this study was a prototype commercial Radarange microwave oven with a power level at 2450 MHz of 2000 watts. The foods were portioned and placed in Pliofilm bags, frozen in a $-12$ F ($-24.4$ C) freezer and stored at 0 F ($-17.7$ C). Four 100-gram lots of vegetables were heated from the frozen state in their Pliofilm bags. The reheating time varied from 3½ to 5 minutes for vegetables and 3¼ to 8 minutes for the meat products. The weight of the meat products varied: 200 grams creamed chicken in a 10 gram rice ring; one piece of chicken breast and one piece of leg with 100 grams of gravy; 150 grams of spaghetti and meat balls; and a 90 gram ham patty.

Dessel et al (1960) compared conventional and microwave cooking on the survival of four bacterial species inoculated into three foods. Microwave cooking was more efficient in destroying *Serratia marcescens*, *Staphylococcus aureus*, *Bacillus cereus* and *Salmonella typhosa* even though the cooking time was one-half to one-ninth as long as the conventional method. The foods used were egg custard, scrambled eggs and ground beef patties.

Grecz et al (1964) found microwave heating more lethal to the spores of *Clostridium sporogenes* than conventional heating. The differences were small when heating to 65 to 70 C (149 to 158 F) and became more pronounced at 85, 95 and 100 C. (185, 203 and 212 F). Delaney at al (1968) using spores of *Aspergillus niger* and an infusion of *Escherichia coli* found no significant difference between conventional heating and microwave heating to 60 C (140 F), when these organisms were inoculated into custard, milk, sucrose or physiological saline.

Studies by Lacey et al (1965) at the Charing Cross Hospital in London, England, in an ultraclean ward where patients were undergoing intensive cytotoxic chemotherapy and were therefore unusually prone to bacterial infection, evaluated the effect of microwave cookery on bacteria in food. The microwave equipment used was also equipped with conventional electric heat and infra red. The study suggested that non-uniform microwave heating could have been responsible for some of the variations in numbers of organisms surviving microwave heating. The conclusion was somewhat unrealistic in that the quantity of the test food used, mashed potatoes inoculated with *Bacillus cereus* spores was sufficient to cover an 8-inch plate to a depth of 1–1½ inches. Such a quantity spread over so large an area practically guaranteed wide temperature variations due to the oven heating pattern. A normal portion of mashed potatoes would have been more meaningful to use. The conclusion reached was that the combination of oven heating, infrared and microwave heating to 160 F (71.1 C) was the most effective in providing sterile or only lightly contaminated foods.

Further support for the effectiveness of microwave reheating in producing bacteriologically safe food came from a study by Woodburn et al (1962). Prepared boned chicken alone, with broth, or white sauce were inoculated with $1 \times 10^6$ bacteria/gram of *Salmonella senftenberg*, *Salmonella typhimurium* or *Staphylococcus aureus*. The product was packaged in polyester-polyethylene laminated plastic bags and heated in a microwave oven for two minutes or immersed in boiling water for 10 minutes. Both heating methods resulted in essentially sterile products and taste panels found little difference in acceptability due to heating method.

In a study by Ockerman et al (1977) meat was inoculated, held for 48 hours at refrigerated temperatures, made into patties and cooked to 85 C (185 F). The conventional cooking time was 12 minutes to reach 140 F (60 C) compared to 24 seconds for microwaves to reach 185 F (85 C). The plate counts before cooking were abnormally high ($1 \times 10^8$) according to these workers and with a less drastic load, microwave cooking might have been adequate. In their summation, Ockerman et al (1977) noted that the short microwave cooking time and the more uniform distribution of temperature throughout the food means that microorganisms are exposed for a shorter time to lethal temperature conditions. Thus, to achieve the same degree of safety as with other methods either the final temperature will have to be increased or the final temperature will have to be maintained for a longer period of time. Two-stage microwave cooking or cooking at lower power for a longer time was mentioned as possible ways to accomplish this.

Crespo and Ockerman (1977) compared low and high temperature cooking with microwave cooking of beef patties. The uncooked samples contained between $24 \times 10^4$ and $80 \times 10^6$ bacteria/gram. The logarithmic reduction as a result of heating method was used for the comparisons rather than total count. Three final temperatures, 34 C (92.2 F), 61 C (141.8 F) and 75 C (167 F), were used in the comparison and log reduction plotted at each temperature. The logarithmic reduction at each temperature was least for microwave heating. Still, at 75 C (167 F) the 3.61 log reduction would place microwave cooked beef within the standards set by the US Army and Skylab programs for count at point of service.

In all studies reported in which foods were inoculated with microorganisms, microwave cooking or heating was shown to be less effective in reducing the plate count than conventional heating methods. This should have been expected since the much shorter microwave exposure means much shorter exposure to lethal temperature conditions as pointed out by Crespo and Ockerman (1977). This situation can be improved somewhat by covering the foods during the microwave heating cycle so that steam generated can be used to increase the thermal load at the surface. This will also reduce the number of cold spots due to microwave heating pattern.

Bunch et al (1976) cooked beef-soy loaves in a convection oven to 60 C and stored them for 24, 48 and 72 hours at $5 \pm 3$ C. Counts in the center of the beef-soy loaves increased substantially during cooling, which took seven hours to reach 7 C. Samples were microwave heated to 80 C. In all cases viable bacteria were found in the center of the microwave heated samples, however, Bunch et al stated that the product was in excellent bacteriological condition.

Dahl et al (1978) studied the effect of initial cooking of beef loaf to different end point temperatures in a convection oven set at $121 \pm 6$ C on the microbiological quality. In cook/chill foodservice, foods are often undercooked to preserve sensory quality if they are to be finished just prior to service. Increasing the initial end point temperature by convection oven cooking resulted in decreasing aerobic mesophilic plate counts, as would be expected. Chilling for 24 hours to 7 C resulted in a further reduction. Terminal heating to 74 C in a microwave oven reduced the mesophilic count still further, as well as aerobic psychrotrophs and coliforms. Viable streptococci were not completely eliminated by the microwave heating step. The end point temperature of 74 C is recommended so that the temperature at point of service to patients is at least 60 C.

Dahl et al (1980a) carried out a study in which the cook/chill foodservice system was simulated in the laboratory. The system involved preparation, initial cooking, chilling, chilled storage, portion and assembly, distribution and microwave heating. Aerobic plate counts were conducted at various stages. A menu of meat loaf, mashed potatoes and green beans was used. The meat loaf was prepared, cooked in a convection oven to an end temperature of 63 to 66 C, drained, packaged and cooled for 24 hours in a walk-in refrigerator. Mashed potatoes were made up from dehydrated potatoes, heated, covered and stored for 24 hours at 6 C. Frozen green beans were transferred to a suitable pan, covered and stored at 6 C. Samples of each, weighing 100 grams apiece, were heated for varying periods of time after they were portioned and held for 2 hours at 6 C to simulate distribution to hospital wards. Meat loaf was plated on paper plates and covered with a plastic dome. Scoops of mashed potatoes weighing 100 grams apiece were placed in foamed plastic cups and lidded with plastic. Green beans were portioned in the same way.

All samples of meat loaf after cooking and chilling met U.S. Army guidelines of $1 \times 10^5$ CFU/g (Colony Forming Units/g) with one exception. A 20 second microwave reheat resulted in slightly higher counts. However reheating for 50, 80 and 110 seconds reduced the counts and 110 seconds reduced the count to less than 30 CFU/gram though the product would be considered unsuitable for serving. The APC of mashed potatoes exceeded U.S. Army or Skylab guidelines even after 85 seconds microwave reheating in one of the three simulations. The APC of green beans met U.S. Army guidelines after 50, 80 and 110 seconds of microwave reheating. Dahl et al (1980a) pointed out that the Aerobic Plate Count of portions of food ranged from $10^1$ to $10^5$ CFU/g after microwave heating but that this could be owing to the range in CFU/g before microwave heating as well as experimental procedure, laboratory and handling errors.

Dahl et al (1980b) studied the effect of microwave finishing on the fate of *Staphylococcus aureus* inoculated into beef loaf, potatoes, and frozen and canned green beans. Cultures of this organism were sprayed onto these foods to give a concentration of $10^7$ CFU/g or more. The experiment was simulated three times. When the end point temperature was 74 to 77 C after microwave heating lethality of *Staphylococcus aureus* was not consistent. In one simulation, after 80 seconds of microwave heating the organism was not detected in beef loaf, whereas 50 seconds of microwave heating resulted in a wide range of counts and could probably be attributed to the uneven distribution of microwave energy during heating.

Under production conditions where strict adherence to sanitary standards is practiced there should be no microbiological problem.

## TEMPERATURE AT POINT OF SERVICE

The Foodservice Sanitation Manual (USDHEW, 1978) states that the temperature of potentially hazardous food reheated for service should be equal to or greater than 74 C (165.2 F). Further, the temperature should be measured in the center of the food. The center temperature is specified, presumably, because by conduction heating methods the center of a food mass is the coldest location. In effect this means that all layers of food around the center have received a much greater thermal load. A simple calculation shows that if the microbiological population is uniformly distributed throughout the food mass most of the organisms will have been exposed to a higher temperature than those close to/or at the center.

By contrast, microwave heating is volume heating, and even though conduction heating also takes place (where temperature differences exist), the conduction heating contribution for small quantities, as in heating plated foods in a foodservice system, is minimal. In addition, if the food is uncovered during heating, evaporative cooling occurs, in effect dissipating heat from the food. Thus, the center of the food mass may not be the coldest location. Geometry, as has been pointed out in an earlier chapter, plays an important role in microwave heating. In some cases the center may be the hottest location, and in the assumption mentioned above these regions would have the least microbial load, essentially representing an overkill situation.

## VARYING RESISTANCE OF MICROORGANISMS TO MICROWAVES

Crespo et al (1977) in beef cooking studies using specific strains of bacteria observed differences in heat sensitivity to cooking methods. *Pseudomonas putrefaciens* was the most sensitive to both cooking techniques [microwave and 176 C (349 F) oven]; *Streptococcus faecalis* was the most resistant to the conventional oven cooking; and *Lactobacillus plantarum* was the most resistant to microwave cooking.

Some researchers have reported on heat activation of spores. Roberts (1968) reported optimal activation of *Clostridium perfringens* spores at 75, 80, 85 and 90 C (167, 177, 185, 194 F). Blanco and Dawson (1974) carried out studies with poultry parts inoculated with *Clostridium perfringens* and cooked by microwaves at 915 and 2450 MHz. Although reduction of spores by microwave energy was minimal (2 log cycles), the procedure provided efficient activation and germination. Browning in 190 C (379 F) vegetable oil resulted in no recovery of cells or spores of this organism.

## SUMMARY

Microwave heating of chilled or frozen prepared food does not guarantee bacteriological safety if the food is heavily contaminated, but it is evident from reports in the literature that a substantial reduction of the bacterial population is effected by this method if heating is continued to a sufficiently high temperature. Cook/chill and cook/freeze microwave foodservice systems should adhere to the following guidelines to insure microbiologically safe foods:

1. Begin with high quality foodstuffs.
2. Use prescribed sanitation practice during preparation
3. Cook to 165 F (73.8 C) or higher.
4. Chill rapidly to 45 F (7.2 C) or less in four hours or less.
5. Package and store at 40 F (4.4 C) or lower, or freeze and store at 0 to −10 F (−17.7 to −23.3 C).
6. Plate foods, cover with a plastic film or plastic banquet cover, hold at 45 F (7.2 C) or lower and microwave heat to 165 F (73.8 C) before serving.

Microwave reheating small food samples such as 100 gram portions, individually, rather than an assembled meal (e.g., 3-components) is somewhat unrealistic, particularly in cook/chill foodservice systems. First of all, the heating rate for individual portions is on the high side and secondly, is less efficient. As shown earlier, when heating small quantities, the actual energy absorbed is much less than the energy available. High heating rates also contribute to uneven heating. The longer reheat time for an assembled meal might result in even lower plate counts. A search of the literature did not reveal any experimental data regarding the microbiology of 3-component meal heating.

Differential heating of chilled meals is less of a problem than with frozen meals. The temperature difference should be much smaller, although a schedule based on heating the protein component to 74 C will almost certainly result in the starch and vegetable components attaining somewhat higher temperatures. A cover over the assembled meal is recommended to contain any steam generated. The steam will condense on the cooler surfaces and improve overall heating uniformity. Without a cover, evaporative cooling of the surfaces will occur.

Some research results noted an increase in plate count during the chilling step. Others reported just the opposite. The size of the food mass being cooled is one explanation offered. In most cases walk-in refrigerators were used that were also being used for chilled storage of other foods. Thus the operating conditions would seem to be subject to considerable variability. The U.S. Army study at Fort Lewis, Washington chilled large quantities of prepared foods in two hours or less in a dedicated liquid Nitrogen cooling system. It is not recommended that all cook/chill foodservice operations adopt liquid Nitrogen cooling, but certainly a dedicated chilling system should be standard practice.

Livingston (1990) speaking on cook/chill foodservice and referring particularly to "sous vide" type operations with specific reference to the potential hazards of psycrotrophic microorganisms, emphasized that even though low temperature refrigeration will extend the shelf-life of such foods, refrigeration is only as reliable as the person monitoring it. To insure abuse resistance and protect against potential growth of *C. botulinum* in the event of refrigeration failure other precautions must be taken. Livingston suggests introducing hurdles to inhibit its growth such as: a pH of 4.5 or lower, controlling water activity to 0.93 or lower, or addition of preservatives.

Those who would enter the foodservice market and/or the consumer market with microwavable chilled foods must recognize the risks to the public health and familiarize themselves with the Hazard Analysis Critical Control Point (HACCP) system. An effective control system is essential to introduction of a line of chilled prepared foods. The article "Application of HACCP to Ready-to-Eat Chilled Foods," by Frank L. Bryan in the July 1990 issue of Food Technology is recommended reading.

O'Donnell (1991) described in some detail the implementation of HACCP by a company nationally marketing chilled entrees. O'Donnell pointed out that the recent outbreaks of food poisoning owing to *Listeria monocytogenes* and *Salmonella* species has raised serious questions about chilled food production practices and resulted in a number of recalls. Outbreaks of *Listeriosis* in England implicated microwave ovens and oven sales dropped dramatically for a while. Actually, microwave ovens were being asked to heat contaminated food that should not have been available in the first place. Nevertheless, the attention given to these outbreaks in the press resulted in the British Ministry of Agriculture, Fisheries and Food being tasked to conduct elaborate studies of microwave oven performance and to develop standards for these ovens that would be reflected on heating instructions on food packages.

# REFERENCES

Baldwin, R.E., Cloninger, M. and Fields, M.L. (1968). Growth and destruction of *Salmonella typhimurium* in egg white foam products cooked by microwaves. Appl. Microbiol. *16* (12), 1929-1934.

Bjorkman, A. and Delphin, K.A. (1966). Sweden's Naaka hospital food system centralizes preparation and distribution. Cornell H & R A Quart. *7*(3):84.

Blanco, J.F. and Dawson, L.E. (1974). Survival of *Clostridium perfringens* on chicken cooked with microwave energy. Poultry Sci. *53*:1823-1830.

Brown, G.H., Hoyler, C.N. and Bierwirth, R.A. (1947). Theory and Application of Radio-Frequency Heating. Van Nostrand-Reinhold, Princeton, New Jersey.

Brown, G.H. and Morrison, W.C. (1954). An exploration of the effects of strong radio frequency fields on microorganisms in aqueous solutions. Food Technol. *8* (8) 361-166.

Bryan, F.L. (1980). Foodborne diseases in the United States associated with meat and poultry. J. Food Prot. *43* (2), 140-150.

Bryan, F.L. (1990). Application of HACCP to ready-to-eat chilled foods. Food Technol. *44* (7) 70, 72, 74-77.

Bunch, W.L., Matthews, M.E. and Marth, E.M. (1976). Hospital chill foodservice systems:

acceptability and microbiological characteristics of beef-soy loaves when processed according to system procedures. J. Food Sci. *41*:1273-1276.

Causey, K.C. and Fenton, F. (1951a). Effect of reheating on palatability, nutritive value and bacterial count of frozen cooked foods. I. Vegetables. J. Am. Dietet. Assn. *27* (5) 390-395.

Causey, K.C. and Fenton, F. (1951b). Effect of reheating on palatability, nutritive value and bacterial count of frozen cooked foods. II. Meat dishes. J. Am. Dietet. Assn. *27* (6) 491-495.

Crespo, F.L, and Ockerman, H.W. (1977). Thermal destruction of microorganisms in meat by microwave and conventional cooking. J. Food Prot. *40* (7) 442-444.

Crespo, F.L., Ockerman, H.W. and Irvin, K.M. (1977). Effect of conventional and microwave heating on *Pseudomonas putrifaciens, Streptococcus faecalis*, and *Lactobacillus plantarum* in meat tissue. J. Food Prot. *40* (9) 558-591.

Dahl, C.A., Matthews, M.E. and Marth, E.H. (1978). Cook/chill foodservice systems: Microbiological quality of beef loaf at five process stages. J. Food Prot. *41* (10) 788-793.

Dahl, C.A., Matthews, M.E. and Marth, E.H. (1980a). Cook/chill foodservice system with a microwave oven: aerobic plate counts from beef loaf, potatoes and frozen green beans. J. Microwave Power *15* (2) 95-105.

Dahl, C.A., Matthews, M.E. and Marth, E.H. (1980b). Fate of *Staphylococcus aureus* in beef loaf, potatoes and frozen and canned green beans after microwave-heating in a simulated cook/chill hospital foodservice system. J. Food Protect. *43* (12) 916-923.

Delaney, E.K., Van Zante, H.J., and Hartman, P.A. (1968). Microbial survival in electronically heated foods. Microwave Energy Appl. Newsl. *1* (3) 11-14.

Dessel, M.M., Bowersox, E.M., and Jeter, W.S. (1960). Bacteria in electronically cooked foods. J. Am. Diet. Assoc. *37* (9), 230-233.

Fleming, H. (1944). Effect of high frequency field on micro-organisms. Electron. Eng. *63* (1), 18-21.

Fung, D.Y.C. and Cunningham, F.E. (1980) Effects of microwaves on microorganisms in Foods. J. Food Prot., *43*, 547.

Genigeorgis, C.A. (1988). Microbiological risk assessment in foods packaged under modified atmospheres. Proc. Pack Alim. '88, March 22-24, San Francisco.

Gray, S.A. (1970a). Method and apparatus for sterilization. U.S. Patent 3,494,722.

Gray, S.A. (1970b). Method and apparatus for controlling microorganisms and enzymes. U.S. Patent 3,494,723.

Gray, S.A. (1970c). Method and apparatus for controlling microorganisms and enzymes. U.S. Patent 3,494,724.

Grecz, N.A., Walker, A.A. and Anellis, A. (1964). Effect of radio frequency energy (2450 MHz) on bacterial spores. Bacteriol. Proc. *1964*, 145.

Ingram, M. and Page, L.J. (1953). The survival of microbes in modulated high frequency voltage field. Proc. Soc. Appl. Bacteriol. *16*, 68-87.

Kossovitsas, C., Navab, M., Chang, C.M. and Livingston, G.E. (1973). A comparison of chilled-holding versus frozen storage on quality and wholesomeness of some prepared foods. J. Food Sci. *38*:901-902.

Lacey, B.A., Winner, H.I., McLellan, M.E. and Bagshawe, K.D. (1965). Effects of microwave cooking on the bacterial counts of food. J. Appl. Bacteriol. *28* (2), 331-335.

Livingston, G.E. (1990). Foodservice: older than Methuselah. Food Technol. *44* (7) 54, 56, 58-59.

Longree, K. (1972). Quantity Food Sanitation, 2nd Ed. John Wiley & Sons, Inc., New York

McGuckian, A.T. (1971). Hospital food service. Microwave Energy Appl. Newsl. *4* (4), 3–6.
Nicholanco, S. and Matthews, M.E. (1978). Quality of beef stew in a hospital chill foodservice system. J. Am. Dietet. Assoc. *72*: 31.
Nyrop, J.E. (1946). A specific effect of high-frequency electric currents on biological objects. Nature (London), *157* (3976), 51.
Ockerman, H.W., Leon Crespo, F., Cahill, V.R., Plimpton, R.F. and Irvin, K.M. (1977). Microorganism survival in meat cooked in microwave ovens. Ohio Rpt. *62* (3), 38–41.
O'Donnell, C.D. (1991) Implementation of HACCP at Orval Kent Food Company, Inc. J. Foodservice Syst. *6* (3) 197–207.
Olsen, C.M., Drake, C.L. and Bunch, S.L. (1966). Some biological effects of microwave energy. J. Microwave Power *1* (2) 45–51.
Proctor, B.E. and Goldblith, S.A. (1951). Electromagnetic radiation fundamentals and their applications in food technology. Adv. Food Res. *3*:120–196.
Roberts, T.A. (1968). Heat and radiation resistance and activation of spores of *C. welchii*. J. Appl. Bacteriol. *31*:133–1444.
Rowley, D.B., Tuomy, J.M. and Westcott, D.E. (1972). Fort Lewis experiment: application of food technology and engineering to central preparation. Tech Rpt. 72-46-FL, US Army Natick Labs., Natick, Massachusetts 01760.
USDHEW (1978). Food service sanitation manual. DHEW Publication No. (FDA) 78-2081, Washington, DC.
Wang, D.I.C. and Goldblith, S.A. (1967). Effects of microwaves on *E. coli* and *B. subtilis*. Appl. Microbiol. *15*, 1371–1375.
Woodburn, M., Bennion, M. and Vail, G.E. (1962). Destruction of *Salmonellae* and *Staphylococci* in precooked poultry products by heat treatment before freezing. Food Technol. *16* (6), 98–100.
Zarotschenzeff, M.T. (1944). Sterilization of meat by induction heating. Quick Frozen Foods *6* (10), 31, 37; *6* (11), 31; *6* (12) 34.

# EPILOGUE

What is the future of microwaves with respect to our eating habits and food product development for the microwave oven user market? It is a big question, and one that requires going way out on a limb to attempt an answer. But some trends or indicators we see and sense provide some degree of confidence in predicting the future.

The microwave oven has a very strong supporter: convenience. Convenience has strong support from microwave oven users. They demand it. Convenience, more than any other advantage of this remarkable device, is the driving force behind microwave oven sales. Energy conservation and waste management have similar strengths with waste management likely to have an edge on energy conservation. Recycling of packaging materials may not have universal appeal, but most likely will have universal compliance. The future may require that we become a society that exists on reheating of prepared foods. Preparing food (meals) from scratch is wasteful of food and energy. We do not need to learn this from space travelers. Yet, perhaps that experience may serve as a valuable education.

Our astronauts know that waste and energy must be diligently controlled. All waste must be stored and returned to earth for disposal. Energy must be carefully meted out or space missions could be in trouble. Space Station Freedom, which is expected to be in orbit before the end of this century, will provide vast quantities of data on life in space with foodservice being one of the important aspects of that life. Eating together will be one of the few social experiences our astronauts will have as a break from their daily routine. A great effort has gone into planning the menus for the space station occupants so that their meals are pleasant and satisfying.

Prepared, portion-packaged foods with optimized nutrient content must be able to provide these nutrients as completely as possible to our astronauts. Only the most gentle reheating method can do that. That method must be convenient, controllable, use the minimum amount of energy, and contribute the least heat and moisture to the space station environmental control system. The packaging must be adequate without being excessive, and it must be compactible after it has served its purpose so as to occupy the minimum storage space until it is returned to Earth. The package must also contribute to the reheating performance in the sense of improving the uniformity of heating. Containers made of renewable resources could be expected to dominate, while polymeric materials that are made from petroleum-based materials will diminish. Aluminum could play a dominant role because it compacts well, requires little energy to remake into foil, and has other advantages as indicated in Chapter 4 on packaging.

In the longer term, that is, the progression to a moon base and perhaps lengthy space missions, the problems become much more difficult. Moon base initially will probably follow the practice of the space station where subsistence on prepared foods shipped periodically from Earth will be the norm. Later, experimental farming will begin and the harvests converted into acceptable edibles. This practice will recycle all waste so that in the manufacture of edibles the trim and non-usable portions will be composted. Moon base will rely more on solar electricity and microwave heating as the most economical and practical usage of energy. Moon base food production concept studies are under way at this time and we can expect to hear reports about this work in the near future. The point to be made here is that space life will include microwave heating as the only practical method to heat food for consumption.

The challenge here on earth will be no less difficult as population growth is factored into the subsistence equation. There are great losses, we are told, in crop harvesting. Some crops spoil rapidly unless processed promptly. The losses, if prevented, could go a long way to feed the undernourished of the world. Processing, storage and distribution techniques add much to the waste stream and this must cease. Energy requirements are tremendous and the cost of energy, particularly for fossil fuels, will increase as they diminish in availability. Eventually, some generations hence, they will have diminished to the point where it will be impractical to extract them from the earth. Long before that happens, saner minds will have prevailed and solar generated electricity and radiation preservation will be commonplace.

Radiation preservation will make it possible to produce very high-quality shelf-stable foods, with quality as good as frozen food. Research, years ago, demonstrated that high quality precooked food products could be produced by irradiation processing. Some of these food products were consumed by astronauts on early orbital and moon landing expeditions. The process is presently being used for preservation of some commodities in a number of countries. All future generations will heat these foods in microwave ovens. A more logical term is microwave devices, since they will not look like ovens and will take little more space than the packages of foods heated in them. The space program will drive the design of these devices.

Food storage will not require refrigeration, though refrigeration may be available for cold beverages and some foods that are more acceptable when chilled. It is possible that our tastes will have changed by that time so that chilling may not be as important. The prepared food containers will be the serving dishes and will be recyclable. Waste will be minimal because foods will be in portion size containers. Penalties might even be imposed for wasting food.

As suggested at the beginning of this epilogue, it is risky to attempt long-range predicting. The retort pouch and irradiated foods were both expected to be commonplace by this time, yet neither has happened. With the success of microwave heating of ready meals it seems unlikely that the retort pouch will be resurrected. Not so with radiation preservation. This process could be adopted, not because it would provide a use for large quantities of radioactive materials, but rather because it can preserve food products, or extend the life of numerous fresh products for the benefit of mankind. In addition because it does not rely on heat for preservation, the quality of such foods should closely approximate the freshly prepared product. The current retorted entrees and ready meals do not seem to have met with consumer acceptance, judging from the space being allocated on supermarket shelves. We will have to wait and see if microwave sterilized entrees and ready meals can succeed. Current research and development efforts in microwave sterilization will have to be increased if this process is to become commonplace in the reasonably near future.

# APPENDIX A

# MANUFACTURERS OF MICROWAVE PROCESSING EQUIPMENT

APV — Magnetronics Ltd.
Stephenson Way, Thetford, Norfolk
IP24 3RP, United Kingdom
Contact: Roger Meredith.
Tel: 44-84 276 2511
Fax: 44-84 276 332

Berstorff Maschinenbau GmbH
P.O. Box 629, D-3000 Hannover 1
Germany
Contact: Klaus Koch
Tel: 49 511 02-0
Fax: 49 511 56 19 16

Berstorff Corporation (USA)
8200-A Arrowridge Blvd,
P.O. Box 240357
Charlotte, NC 28224
Contact; Wilfried Schlegel
Tel: 704 523-2614
Fax: 704 523-4353

Cober Electronics, Inc.
102 Hamilton Avenue
Stamford, CT 06902
Contact: Bernard Krieger
Tel: 203 327 0003
Fax: 203 359 6319

ETC Process
86-106 Avenue Louis Roche
C 205
92230 Gennevilliers, France
Contact: M.C. Brugalieres
Tel: 33 1 40 85 17 00

IMI — Industrial Microwaves
Route de rambouillet
F-78680 Epone
France
Tel: 33-1-30-95 66 76
Fax: 33-1-30-95 33 87

Microdry Inc.
7450 Highway 329
Crestwood, KY 40014
Contact: John Wiedersatz
Tel: 502 241 8933

MicroHeat Sweden AB
P.O. Box 7097
S-172 07 Sundbyberg, Sweden
Contact: Benny Berggren
Tel: 46 8 733 95 10
Fax: 46 8 98 73 51

Micro-Onde Energie Systemes
2-4 Avenue de la Cerisaie
Platanes 307
94266 Fresnes Cedex, France
Contact: M.A.-J. Bertaud
Tel: 33 1 46 68 39 39

Microwave Heating Ltd.
Unit 2, Heron Trading Estate
Sundon Park, Luton, Beds.
LU3 3BB, United Kingdom
Contact: Ralph Shute
Tel/Fax: 44 58 258 4747

OMAC Impianti s.r.l.
via Industria 6
42019 Pratissolo di Scandiano
Reggio Emilia, Italy
Contact: Dr. Eng. Carlo Coluccio
Tel: 39 2921 73 098
Fax: 39 2921 52 477

OMAC Representative, USA
T.W. Kutter, Inc.
279 D Center St
Holbrook, MA 02343
Contact: Linda Harlfinger

Oshikiri Machinery Ltd.
14-12, 4-Chome
Ohmori-Nishi, Ohta-Ku
Tokyo, Japan
Tel: 81 3 761 9171

Raytheon Company
Industrial Equipment
Foundry Avenue
Waltham, MA 02154

Contact: Ronald Snider
Tel: 617 642 4244
Fax: 617 642 3718

SAIREM
24 rue Louis Saillant
69120 Vaulx en Velin, France
Contact: M. Jean-Paul Bernard
Tel: 33 72 04 09 25

Schrade Hochfrequenztechnik GmbH
Ferdinand Gabriel Weg 12
4770 Soest, Germany
Contact: Frank Wisgalle
Tel: 49 292 12 24 77

SFAMO
69653 Villefranche/Saone
France
Contact: Robert Bellavoine

# INDEX

Advantages
  of microwave ovens, 12, 17
  of packaging, 96
Aluminum, 42, 54
Appliances for use in microwave ovens, 38
Athermal effect on bacteria, 190
Ascorbic acid, 170-171
*Aspergillus niger*, 194

*Bacillus cereus*, 194
*Bacillus megatherium*, 193
Batters
  modified corn starch, 133
  soft-white wheat flour, 133
  waxy maize starch, 133
  yellow corn flour, 133
Browning, 5, 7, 117-128, 130, 158
  devices, 120
    browning dish, 120
    chemical, 122
    ferrite devices, 120
    susceptors, 121
  formulations, 117-120
    meat, 118
    pastry, 119
  Maillard reaction, 149
  ovens, 124

Carotene, 168
Catalase, 166
Chlorophyll, 171
*Clostridium sporogenes*, 194
Coatings
  alginate, 134
  carboxy methyl cellulose, 134
  hydroxy propyl methyl cellulose, 134
  methyl cellulose, 134
  starch, 134
Computer simulation, 53
Containers
  bowls, 138
    coextruded, 138
  dishes
    baking, 132
    browning dishes, 131
    polyester, 146
  pouches, 154, 158
  trays, 95
Container materials
  aluminum, 42, 54
  ceramic, 42, 89
  ferrites, 39-41
  glass-ceramics, 88
  paper, 94
  plastics, 89-91
    phenylene sulfide, 42
    polycarbonate, 38
    polyester, 42,
    polysulfone, 42
    Teflon, 42
  stainless steel, 42
  stoneware, 42
C-values, 156

Dielectric properties of foods, 50
Differential heating, 146

Emission, 70
  monitors, 80

performance standards, 80, 199
PL 90-602, 80
Exposure standards, 81
Evaporative cooling, 52, 61

Food products
  brownie mixes, 132
  cake mixes, 132
  chicken nuggets, 26
  crepes, 131
  entrees, precooked, 19
  French toast, 131
  meals
    chilled, 157, 189
    frozen, 16, 129, 131, 154
    precooked, 16, 22, 131
    ready, 112, 157
    shelf-stable, 189
  pancakes, 36, 131, 136
  pastries, 131
  pizza, 26, 36, 121, 130
  popcorn, 36, 99–100, 102–105, 137
  potatoes,
    French fried, 24, 26, 122, 125, 134
  sandwiches
    hamburger, 129, 141
    submarine, 129
  snacks, 139, 140, 160
    coated, 140
    dry fruit chip, 141
    high protein, 140
    potato chips, 140
    puffable, 140
  soups, 131, 137, 139
  thick shakes, 15
Foodservice systems, 8
  cafeteria, 18
  cook/chill, 191–192, 196
    advantages of, 192
    guidelines, 198
  cook/freeze
    guidelines, 198
  Gingham Kitchen, 14
  heating guidelines, 31

hospital, 15, 31
  A.G.S. system, 191
  Kaiser Foundation, 16
  NAAKA system, 191
  Walter Reed Army, 16
  West Jersey, 17
    advantages, 17–18
in-flight, 26–27
in-plant, 18
military, 28
MUST, 30
railroad, 30
restaurant, 9–15
  advantages, 12
school lunch, 18–19
shipboard, 30
snack bar, 14
SPEED Bakery, 30
SPEED Kitchen, 28
vending, 20, 22
  sandwiches, 20
Fundamentals of microwave heating, 47–60
Future microwave ovens, 82–84
F-values, 156

Geometry, 53, 145

Half-power depth, 50
History of the microwave oven, 1–46

*Lactobacillus plantarum*, 197
*Listeria monocytogenes*, 199

Maillard reaction (*See* Browning)
Microbiology, 154, 156
  standards, 159
Microwave baking, 132
Microwave cooking
  effect on
    *Bacillus cereus*, 194
    *Lactobacillus plantarum*, 197
    *Pseudomonas putrifaciens*, 197
    *Salmonella typhosa*, 194

*Serratia marcescens*, 194
*Staphylococcus aureus*, 194
*Streptococcus faecalis*, 197
Microwave energy
  effect on enzymes
    catalase, 166
    peroxidase, 166
  effect on fat, 180, 183
  effect on microorganisms, 189–199
    athermal effect, 190
    thermal effect, 190
  effect on moisture, 180–184
  effect on palatability, 180
  effect on vitamins, 166–185
    ascorbic acid
      in vegetables, 168, 170–171, 176–177
    carotene, 168
      in vegetables, 168
    riboflavin
      in cake mixes, 167
      in meats, 166, 172–173
      in vegetables, 169, 177
    thiamine
      in cake mixes, 167
      in meats, 166, 172–175, 181, 183
      in prepared foods, 177
Microwave heating
  activation of spores of
    *Clostridium perfringens*, 197
  characteristics of, 48
  effect on
    *Aspergillus niger*, 194
    *Bacillus cereus*, 193–194
    *Bacillus megatherium*, 193
    *Clostridium sporogenes*, 194
    *Escherichia coli*, 194
    *Salmonella senftenberg*, 194
    *Salmonella typhimurium*, 194
    *Staphylococcus*, 193, 196
  efficiency, 59
  fundamentals, 47
  ionic conduction, 48
  pattern, 68
    determination, 68
    methods, 68–69
  physical factors
    density, 52, 59–60
    evaporative cooling, 51, 61–62
    final temperature, 61
    geometry, 52–54
    initial temperature, 61
    specific heat, 51, 57–58
    surface to volume ratio, 51, 56
    thermal conductivity, 51, 61
  polar molecules, 48
Microwave leakage (*See* Emission)
Microwave oven
  accessories, 38
    cookbooks, 38
    cookware, 38
    sales of, 38
  advantages, 12, 17
  appliances, 38
    active, 38
      baker, 41–42
      cooker, 41–42
      corn popper, 38
      griddle, 38–40
      steamer, 38–39
    passive, 38
      bacon racks, 38
      corn popper, 38
      egg pans, 38
      muffin pans, 38
      roasting racks, 38
      tube pans, 38
      turntables, 38
  basic components, 67
    cavity, 67
    controls, 67, 73
    door, 67, 70
      seal, 70
      choke, 70
    magnetron, 67, 71
    mode stirrer, 67, 72
    power supply, 67, 72

waveguide, 67, 71
commercial, 2, 26
compact, 37
consumer, 31, 78–79, 82–83
continuous, 17
convection, forced 7, 77
conveyorized, 17, 148
design factors affecting microwave heating
  cavity material, 63
  cavity volume, 63
  feed system, 63
  field pattern, 63
  floor material, 64
  line voltage, 64
  magnetron age, 64
  power, 62
  power supply, 64
  time base, 64
features, 35
  bar code reader, 146
  fuzzy logic, 76
  gas sensing, 74
  humidity sensing, 36, 74
  programmable, 36
  recipe cards, 74
  recognition devices, 36
  speaking, 36, 74
  temperature sensing, 75
  turntable, 36
  variable power, 75
  weight sensing, 76
impingement, 77
MUST, 30
pattern, 5, 68–69, 155
safety, 79
sales, 2, 36, 157
SPEED, 28
standardization, 79
thawing, 8
vending, 20, 24
  prototype, 21
Microwave processing

baking, 166
  effect on
    riboflavin, 166–167
    thiamine, 166–167
blanching, 166–167
  effect on
    ascorbic acid, 166, 171
    catalase, 166
    chlorophyll, 171
    peroxidase, 166, 171
high temperature-short time, 148, 151
in-package
  pasta, 189
  sliced bread, 189
  sterilization, 91, 112, 148, 151–152, 156, 159
pasteurization, 91, 158
  effect on
    *Clostridium perfringens*, 191–192
    *Salmonella*, 191–192
tempering, 131
vacuum drying, 141
Microwave properties
  attenuation constant, 51
  dielectric conductivity, 49
  dielectric constant, 50
  dielectric loss factor, 50
  field strength, 50
  frequency, 50
  half-power depth, 50–51
  loss tangent, 50
  penetration, 50
Multitherm, 156

Nonthermal effects, 190
Nutrition, 165

Packaging, 87–116
  advantages, 96
  design, 158

dual-ovenable, 93, 95, 129
functions, 87
control heating, 99
materials
  glass, 87–88, 110
  metals, 87, 95–99
    aluminum foil, 92–93, 112, 129, 144
  paper, 87, 94
    molded pulp, 94
  paperboard
    PE coated, 19, 94
  patents, 99–112
  plastics, 87, 89
    acrylonitrile, 90
    barrier films, 91–92, 155
      coextruded films, 92
      crystalline PET, 92
      ethyl vinyl alcohol, 92
      glass coated polyester, 92
      polyvinylidene chloride, 92
      thermoformed, 92
    butadiene styrene, 90
    effect on flavor, 93
    filled polyester, 90–91
    nylon, 90
    poly butylene terephthalate, 90
    polycarbonate, 90, 93
    polyethylene, 90–91
    poly(methylpentene), 90
    poly (phenylene oxide), 90
    polypropylene, 90
    polystyrene, 91
    polysulfone, 90
    styrene acrylonitrile, 90
  modified atmosphere, 92, 189
  retortable containers, 92
  shelf-stable, 92
Peroxidase, 166
Phantom food, 53
Positional relationships, 55
Product development, 129–163
*Pseudomonas putrifaciens*, 197

Quality of microwave heated food, 156
Quality control, 159

Riboflavin, 167
Runaway heating, 145

*Salmonella senftenberg*, 194
*Salmonella typhimurium*, 194
*Salmonella typhosa*, 194
Seminar, microwave, 5
*Serratia marcescens*, 194
Shadowing, 55
Shielding, 54, 100, 105–110, 129–131, 142–143, 158
Space charge polarization, 49
Specific heat, 52, 55, 145
Standards
  exposure, 81–82
  performance, 80
*Staphylococcus aureus*, 194
*Streptococcus faecalis*, 197
Surface to volume ratio, 52, 55
Susceptors, 94, 111, 135, 143
Symbol
  microwavable, 157

Temperature measurement, 149
Temperature probes, 73–74
Thermal conductivity, 52, 60, 152
Thermography, 53
Thermometry
  paper strip, 149
Thiamine, 167

Vending
  food, 20–26, 129
  machines, 20, 135, 145
  sandwiches, 20

Waste management, 144
Waveguide, 67, 71
Weight sensing, 76